DNA Sequencing - History, Present and Future

Edited by Ibrokhim Y. Abdurakhmonov

Published in London, United Kingdom

DNA Sequencing - History, Present and Future
http://dx.doi.org/10.5772/intechopen.1003355
Edited by Ibrokhim Y. Abdurakhmonov

Contributors
Alexandra Constantinescu, Amit Mathews, Andreea Mirela Caragea, Ichiro Fujihara, Ileana Constantinescu, Laurentiu Camil Bohiltea, Mevlüt Arslan, Mitsuru Furusawa, Motohiro Akashi, Radu-Ioan Ursu, Rhitisha Sood, Vivek Singh

First published in London, United Kingdom, 2025 by IntechOpen
IntechOpen is the global imprint of INTECHOPEN LIMITED, registered in England and Wales, registration number: 11086078, 167-169 Great Portland Street, London, W1W 5PF, United Kingdom

For EU product safety concerns: IN TECH d.o.o., Prolaz Marije Krucifikse Kozulić 3, 51000 Rijeka, Croatia, info@intechopen.com or visit our website at intechopen.com.

British Library Cataloguing-in-Publication Data
A catalogue record for this book is available from the British Library

DNA Sequencing - History, Present and Future
Edited by Ibrokhim Y. Abdurakhmonov
p. cm.

This title is part of the Genetics Book Series, Volume 5
Topic: Genomics
Series Editor: Kenji Ikehara
Topic Editor: Irina Vlasova-St. Louis

Print ISBN 978-1-83634-287-8
Online ISBN 978-1-83634-286-1
eBook (PDF) ISBN 978-1-83634-288-5
ISSN 3049-7094

If disposing of this product, please recycle the paper responsibly.

IntechOpen Book Series
Genetics
Volume 5

Aims and Scope of the Series

"Genetics," which has been proud of its tradition since Mendel presented his research results in 1865, initially progressed quite slowly due to simple observational approaches of individuals and groups. However, the discovery of double-stranded DNA by Watson and Crick about 70 years ago triggered rapid progress in life sciences, including genetics, which was primarily conducted using Escherichia coli and bacteriophages infecting E. coli. Subsequently, genetics has achieved remarkable developments, such as understanding genetic disorders, including cancers, through research on the biogenesis and differentiation of plants and animals.
The two topics of this book series - Human Genetics, and Genomics - will address important areas of advancement in genetics.
Human Genetics: After fundamental genetics, initially studied with the main goal of revealing the functions of individual genes and proteins, genetics expanded from understanding the genetic system itself to understanding many infectious diseases caused by bacteria and viruses. Consequently, human beings are now overcoming infectious diseases by developing medicinal chemicals, including antibiotics and vaccines. However, genetic disorders remain challenging to cure up to now. Nevertheless, even the cure for them, including various cancers, is coming closer to reality due to the rapid progress of human genetics. In this way, the welfare of human life continues to improve, and even longevity, which was once a dream, has been achieved to some extent in recent years.
Genomics: On the other hand, the understanding of the comprehensive interrelationship of whole genes or whole proteins functioning in one organism has become possible now, as research has entered the era of genomics, owing to the rapid progress of base sequence analysis and bioinformatics. The development of genomics has further made it possible to understand the evolutionary processes of organisms through comparative studies among the genomes of many organisms.
This book series will discuss the findings obtained during the advancement of human genetics and genomics. It is also expected that this series will trigger the formation of a better world composed of human beings and all other organisms on Earth through discussions of research results obtained under the development of general genetics.

Meet the Series Editor

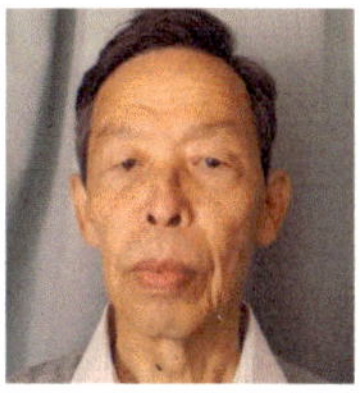

Kenji Ikehara graduated from the Department of Industrial Chemistry, Faculty of Engineering, Kyoto University in 1968. He received his B. Eng. (1968) and subsequently earned M. Eng. (1970) and D. Eng. (1976) degrees from Kyoto University. He began his career as a research associate in the Faculty of Science at the University of Tokyo before moving on to become an associate professor in the Faculty of Science at Nara Women's University. He was later promoted to professor and subsequently served as the dean of the Faculty of Science at Nara Women's University. Additionally, he held the position of director at the Nara Study Center of the Open University of Japan. For approximately 15 years, he focused his research on sporulation initiation of Bacillus subtilis. Later, he shifted his focus to the origins and evolutionary processes of microbial genes, the genetic code, proteins, and life. He has proposed several hypotheses, including the GC-NSF(a) hypothesis on the origin of genes, the GNC-SNS hypothesis on the genetic code, the protein 0th-order structure hypothesis on the origin of proteins, and the [GADV]-protein world hypothesis (GADV hypothesis) on the origin of life. Furthermore, he served as the local chair of the International Conference, Origin 2014, held in Nara in 2014.

Meet the Volume Editor

Ibrokhim Y. Abdurakhmonov received his BS in Biotechnology (1997) from the National University in Uzbekistan, an MS in Plant Breeding (2001) from Texas A&M University in the USA, and a Ph.D. in Molecular Genetics (2002) and DSc in Genetics (2009) from the Academy of Sciences of Uzbekistan. He founded the Center of Genomics and Bioinformatics of Uzbekistan in 2012. He received the 2010 TWAS prize and ICAC Cotton Researcher of the Year 2013 for his outstanding contribution to cotton genomics and biotechnology. Dr. Abdurakhmonov was elected as a fellow of The World Academy of Sciences (TWAS) in 2014 and a member of the Academy of Sciences of Uzbekistan in 2017. In 2023, he was awarded the UNESCO-Guinea Equatorial Life Science award for contributing to molecular marker development and disease resistance in cotton.

Contents

Preface

DNA sequencing technologies, a groundbreaking opportunity for the rapid advancement of biological science and research, have been a major driving force in life science research development, leading to an understanding of life. They have provided novel discoveries and high-throughput medical, agriculture, and forensic science solutions. The foundation of DNA sequencing traces back to the Sanger sequencing technique, marking a pivotal milestone in reading the structure of DNA. Subsequently, the advent of automated sequencers and high-throughput sequencing platforms, known as Next-Generation Sequencing (NGS), catalyzed a paradigm shift in genomics. This shift enabled us to decode the whole genome of living matters with unprecedented speed and accuracy, ushering in a new era of genomics and bioinformatics. This new era offers a fresh perspective on life, from personalized medicine and disease diagnostics to evolutionary studies and biodiversity conservation.

While the advancement of next-generation sequencing technologies has already provided an unprecedented, precise understanding of genome complexities, the future holds even more promise. Advances in single-molecule sequencing and nanopore technology are on the horizon, offering exciting possibilities for humanity. Emerging platforms of real-time, portable, affordable DNA sequencing tools will undoubtedly revolutionize fieldwork and point-of-care diagnostics, heralding a new era of genomic medicine, agriculture, and food security.

In this book, *DNA Sequencing – History, Present and Future*, we have compiled updated and diverse overviews of the international genomics research community. The journey through the history, present, and future of DNA sequencing describes innovation, discovery, and limitless possibilities in front of us, deepening our exploration. Chapters have highlighted the present power and transformative potential of DNA sequencing, providing humanity with a tool for understanding, feeding, healing, and exploring the genetic blueprint of the Tree of Life. Readers will enjoy going through the developmental stages of DNA sequencing methodologies and the availability of technological platforms with examples of applications advancing our knowledge in understanding genetic processes. The authors also addressed future developments in DNA sequencing and massive data mining so humanity can gain a brighter, safer, healthier future than before with less poverty and hunger.

I am sure that all chapters in this book will be helpful for life science students, researchers, and interested readers.

I thank all the authors of the book chapters for their efforts and invaluable contributions to this book. I am also thankful to the IntechOpen book department and publication managers for the opportunity to work on this book project and help with my editorial duties.

Ibrokhim Y. Abdurakhmonov
Center of Genomics and Bioinformatics,
Academy of Sciences of Uzbekistan,
Tashkent, Uzbekistan

Section 1

Development of DNA Sequencing Methods and Tools

Chapter 1

DNA Sequencing: A Brief History

Amit Mathews

Abstract

DNA sequencing, which deciphers DNA nucleotide sequence, has had a transformative impact on biology and medicine. The quest began in 1953 with Watson and Crick's discovery of the DNA double helix, based on Franklin and Wilkins' pioneering work in X-ray crystallography. Arthur Kornberg's 1956 discovery of DNA polymerase further advanced the field, laying the groundwork for future sequencing technologies. In 1977, Frederick Sanger's chain-termination method (Sanger sequencing) and Maxam-Gilbert's chemical sequencing emerged as the first viable techniques for reading DNA. Sanger sequencing, in particular, remains widely used for analyzing shorter DNA sequences today. The 1980s saw the introduction of automated sequencers, which dramatically boosted throughput, precision, and accessibility. Launched in 1990, The Human Genome Project (HGP) utilized Sanger sequencing to map the human genome, marking a historic achievement that is often compared to landing on the moon. The early 2000s ushered in next-generation sequencing (NGS) technologies, such as pyrosequencing, sequencing by synthesis (SBS), and SOLiD, offering faster and more cost-effective sequencing. Today, third-generation technologies like Oxford Nanopore's "nanopore" sequencing and PacBio's Single-Molecule Real-Time (SMRT) sequencing enable longer reads and real-time data. These advancements have revolutionized genomics, driving progress in evolutionary biology, precision medicine, and forensic science.

Keywords: DNA sequence, next-generation sequencing, Sanger sequencing, human genome project, sequencing by synthesis, genomics

1. Introduction

1.1 DNA sequencing

A fundamental aspect of contemporary biology and medicine is DNA sequencing, which is the process of deciphering the exact nucleotide order within a DNA molecule. It has completely transformed the fields of Biology and Medicine by providing molecular insights into Genetics and the more recent field of Genomics. Its importance cannot be overstated because it is the foundation for numerous advancements in genetics, genomics, and personalized medicine. Scientists can discover the underlying patterns governing the growth, operation, and reproduction of all living things by decrypting the genetic information encoded in DNA. The ability to read genetic code has revolutionized our understanding of the biological realm, resulting in pioneering findings in several domains including pathology, evolutionary biology, and biotechnology [1]. DNA sequencing has entirely changed medical diagnosis and therapy approaches. Targeted medicines have been developed as a result of the capacity to identify the genetic abnormalities causing diseases, providing patients with previously incurable disorders with new hope. For example, in cancer research, sequencing technology has made it possible to identify particular mutations within tumors, has made it possible to create individualized treatment programs that are specific to each patient's genetic profile. DNA sequencing is essential to the science of pharmacogenomics because it helps explain how specific genetic variations affect how each person responds to medication, which can result in safer and more effective therapies. Apart from its influence on medicine, DNA sequencing has led to tremendous progress in the fields of environmental studies, forensic science, and agriculture. It has made it easier to develop genetically modified agriculture crops that have improved characteristics including tolerance to diseases, pests, and environmental challenges. DNA sequencing is also a forensic science technique that analyses genetic material recovered from crime scenes to provide vital evidence for case resolution. Through metagenomics, sequencing technologies have improved environmental studies and provided fresh perspectives on biodiversity and ecosystem processes [2].

1.2 Evolution of DNA sequencing

The development and improvement of DNA sequencing technology has been astonishing. The origins of DNA sequencing technology can be traced back to the discovery of the DNA double helix structure by James Watson and Francis Crick in 1953 [3], which in turn was based on X ray crystallography studies by Rosalind Franklin and Maurice Wilkins [4]. In 1956, Arthur Kornberg's identification of DNA polymerase—the enzyme behind DNA replication emerged as the critical tool for DNA sequencing. From its modest beginnings in the 1970s, the field advanced rapidly with the invention of Maxam Gilbert's chemical sequencing and Frederick Sanger's chain-termination sequencing, which further propelled the search for quicker, less expensive, and more precise sequencing techniques [5]. Frederick Sanger's original approach, which established the field, was based on the chain-termination methodology and gained widespread prominence due to its simplicity and reliability. Numerous early genetic discoveries, such as the first full sequencing of a viral genome and the later, enormous Human Genome Project, were made possible thanks mainly to Sanger sequencing [5].

As the market for high-throughput sequencing expanded, it became clear that Sanger sequencing had drawbacks, most notably its high cost and low throughput. This resulted in a dramatic shift in the area with the introduction of next-generation sequencing (NGS) technology in the mid-2000s [2, 5]. Illumina and other NGS technology made it possible to sequence millions of DNA fragments in parallel, which significantly cut down on sequencing time and expense. Personalized treatment and large-scale genomic studies are now possible chiefly because of these developments. Third-generation sequencing technologies, such as those created by Oxford Nanopore Technologies and Pacific Biosciences, have expanded the realm of possibilities in recent years. These technologies make real-time single-molecule sequencing possible by enabling greater read lengths and direct detection of epigenetic changes [6]. DNA sequencing is still evolving today, with new technologies promising to improve our understanding and manipulation of the genetic code and maybe lead to new discoveries and applications in research, medicine, and other fields.

2. Early milestones in DNA sequencing: A historical perspective

2.1 Discovery of DNA structure

The discovery of the DNA structure itself is closely intertwined with the early history of DNA sequencing. In the middle of the nineteenth century, Gregor Mendel established the foundational principles of inheritance through his research on pea plants, paving the way for our current understanding of the molecular basis of life. Proteins were long suspected to be the carriers of genetic information throughout the nineteenth century, although the mechanism was completely unknown. It was not until the early twentieth century that scientists started to suspect that it could be DNA, and not proteins that was the magic molecule carrying precious genetic information. The 1944 tests by Avery, MacLeod, and McCarty, which showed that DNA could transfer genetic features between bacteria, substantially verified this notion. These discoveries laid the groundwork for the subsequent discovery of the structure of DNA, which would transform biology and enable DNA sequencing [7].

The breakthrough double-helix model of DNA was postulated by James Watson and Francis Crick in 1953, and was based on the research of Rosalind Franklin, Maurice Wilkins, and other scientists. This was one of the most important discoveries in Biology and a crowning achievement [3]. This finding was made possible mainly by Franklin's X-ray diffraction photographs of DNA, which offered crucial proof of the double helical structure of DNA [4]. Using the two sugar-phosphate backbones pointing in opposing directions and joined by pairs of nitrogenous bases, adenine (A), thymine (T), guanine (G) and cytosine (C), Watson and Crick's model described DNA as a double-stranded helix. The base-paring mechanism explained how DNA could replicate itself and the mechanism of genetic information storage and transmission from one generation to the next was suddenly revealed. Watson and Crick famously noted this in their paper in *Nature* stating: "It has not escaped our notice that the specific pairing that we have postulated immediately suggests a possible copying mechanism for the genetic material [3, 4, 7]." The identification of DNA's double helix structure was a paradigm-shifting piece of work. It offered the molecular foundation for comprehending evolution, heredity, and genetics. The advancement of molecular biology was greatly aided by the simple process for copying genetic information that Watson and Crick's model proposed. This finding had far-reaching consequences creating whole new avenues for study in the quest to understand the genetic code [3, 4, 7].

Once the structure of DNA was known, researchers turned their attention to the reading and interpretation of this genetic data. One of the most important discoveries was understanding that the nucleotides in the DNA strand code for instructions that create proteins and other molecules necessary for life. As a result, James Watson and Francis Crick developed the central dogma of molecular biology, according to which genetic information moves from DNA to RNA to protein. However, to truly comprehend how genetic information was encoded, techniques for determining the precise sequence of these bases had to be developed, a task that would keep scientists busy for the next few decades. The discovery of the structure of DNA has both immediate and long-term implications for the field of genetics. It offered a concrete framework for understanding mutations, which are alterations in the DNA sequence that may affect the composition and functionality of proteins and, eventually, an organism's characteristics. The discovery that mutations are variations in the DNA sequence paved the way for the eventual development of DNA sequencing technologies, which let researchers detect and analyze these variations at the molecular level [3, 4, 7].

2.2 Initial sequencing efforts

Following the discovery of the structure of DNA, the next major challenge for scientists was to create methods for sequencing DNA which meant decoding the precise arrangement of nucleotide bases on a strand of DNA. Simpler molecules like RNA were the focus of early sequencing attempts, since they were smaller and more accessible to reach than DNA. The successful sequencing of the first complete nucleic acid, the yeast alanine transfer RNA (tRNA), by Robert Holley and colleagues in 1965 was one of the earliest important turning points in the field of nucleic acid sequencing [8]. This was a ground-breaking achievement, since it showed that it was possible to determine the RNA nucleotide sequence, which is closely related to DNA. In Holley's work, the RNA molecule was partially hydrolysed, the fragments were separated, and their sequences were examined. This laborious and meticulous procedure offered a blueprint for the eventual sequencing of more complicated nucleic acids [8]. Although Holley worked mostly with RNA, attempts were also being made to create techniques for sequencing DNA. Using chemical techniques to selectively break down DNA at particular bases was one of the first ways to sequence DNA. By examining the resultant fragments, scientists were able to assemble the original sequence as it existed in nature. This method was pioneered by Ray Wu and his colleagues in the 1970s, who used exonucleases to gradually break down DNA molecules from their ends and then identify the terminal nucleotide at each stage. Although Wu's method provided some information about the DNA sequence, it had limited ability to sequence long stretches of DNA accurately [9].

The Maxam-Gilbert and Sanger procedures, which were developed in the late 1970s, were the first true breakthrough in DNA sequencing. Chemical sequencing, or the Maxam-Gilbert method, was created at Harvard University by Allan Maxam and Walter Gilbert [10]. This technique involves chemically modifying specific bases of DNA followed by cleaving DNA at the modified locations. The pattern of bands produced by running the resultant fragments on a gel could be used to determine the DNA's sequence. Although the Maxam-Gilbert method was a significant advancement, it was challenging to scale for longer sequences and involved radioactive labelling, which made it hazardous [10]. The more resilient approach was the Sanger method, created in 1977 by Frederick Sanger and associates, which established the industry standard for DNA sequencing. Sanger's technique, called the chain-termination method, involved synthesizing a new DNA strand complementary to the template strand and adding dideoxynucleotides (ddNTPs) to force the synthesis to terminate at specific locations. Using a mixture of standard nucleotides and labeled ddNTPs, Sanger generated a series of DNA fragments of different lengths, each terminating at a particular nucleotide. After sorting these fragments based on size using gel electrophoresis, the original sequence of the DNA molecule could be deduced from the banding pattern [5]. The Sanger method was more accurate, efficient, and easier to automate than earlier approaches, making it a major advancement. It was soon the preferred technique for sequencing DNA, and in 1977 it was used to sequence the first complete genome of a virus (bacteriophage φX174). This accomplishment established the foundation for extensive sequencing initiatives, such as the Human Genome Project, and proved the efficacy of the Sanger method [5, 11].

Launched in 1990, the Human Genome Project (HGP) was a monumental worldwide endeavor to sequence all the approximately 3 billion base pairs of the human genome. The HGP was launched with such great fanfare that this laid the groundwork for the next level of large-scale improvements to the technology. HGP was wildly ambitious and brought together scientists from all parts of the world united in the spirit of science [11]. It has since become one of the greatest scientific triumphs of the twentieth century and the completion of the first draft of the human genome in 2003 was one of the most widely celebrated scientific achievements of the new millennium. Deciphering human genome turned into the ultimate Molecular Biology project. This was the first time we read our own blueprint, and DNA sequencing technology was right in the middle of all this action. This also captured popular imagination at the time. Managed by the National Institutes of Health (NIH) and the Department of Energy (DOE) in the United States, and involving contributions from labs around the world, the HGP used automated Sanger sequencing and successfully charted all the genes of the human genome, which consists of over 3 billion base pairs, establishing a baseline genome for future research. Such was the impact of HGP, it is often called the biological equivalent of the moon landing. The rough draft was ready by 2003 and over 2.7 billion US dollars were spent to achieve the feat [11]. The successful sequencing of the human genome paved the way for the advancement of next-generation sequencing technology and opened up new options for genetics, medicine, and biology. The early days of DNA sequencing are marked by tenacity, resourcefulness, and teamwork. Understanding of the molecular foundation of life has advanced significantly thanks to scientific discoveries such as the structure of DNA and the creation of the first sequencing techniques. These initial initiatives set the stage for the swift developments in sequencing technology, revolutionizing our capacity to examine and work with the genetic code [11].

Looking back on the early history of DNA sequencing, it is evident that Watson and Crick's discovery of the DNA structure was a turnkey moment in history that created the pathway for all later advances in the area. Their double helix discovery motivated a generation of scientists to study the molecular mechanics underlying life. It also provided the key to understanding how genetic information is stored and transmitted. Large-scale sequencing studies like the Human Genome Project would eventually be made possible by the methods and tools developed by the early sequencing efforts, which, despite being labour-intensive and sluggish, proved that it was possible to read the genetic code [7–11]. It was a long and challenging road from the discovery of the DNA structure to the first successful sequencing of an entire genome, but one that revolutionized our understanding of biology and created new avenues for study, treatment, and technology. DNA sequencing continues to spur innovation in various domains, including synthetic biology and personalized medicine and is now a standard technique in laboratories worldwide. The groundwork for this revolution was built by the early proponents of DNA sequencing, including Watson, Crick, Sanger, and their associates. Their contributions still echo in the scientific community [3–5, 10, 11].

3. First-generation sequencing

3.1 Maxam-Gilbert sequencing

One of the earliest methods for sequencing DNA was Maxam-Gilbert sequencing, created in 1977 by Allan Maxam and Walter Gilbert [10]. This approach marks a critical turning point in the field of molecular biology. This technique—also called chemical sequencing—marked a significant development genetics and genomics by enabling the direct reading of DNA nucleotide sequences [10, 12]. The Maxam-Gilbert method used chemical reactions to break DNA strands at particular nucleotides, which was an innovative approach. The first step in the procedure is to radioactively mark one end of the DNA strand. Then, four different chemical treatments are applied to the labeled DNA, each of which targets a different nucleotide: guanine (G), adenine (A), cytosine (C), or cytosine/thymine (C/T). Breaks at the locations matching the targeted nucleotides are then mapped out. After chemical cleavage, polyacrylamide gel electrophoresis is used to sort the DNA fragments based on size. When X-ray film is exposed to the gel, an autoradiogram is created that shows bands corresponding to the DNA fragments' lengths. Researchers determined the DNA sequence by comparing the band patterns from the four reactions, where each lane represented the fragments produced by a different chemical treatment based on the four nucleotides [10, 12].

One of the main benefits of Maxam-Gilbert sequencing was its precision in sequencing tiny DNA fragments. It was beneficial for examining DNA sections, that were difficult to clone, including highly repetitive or G-C rich regions. Although Maxam-Gilbert sequencing was an ingenious method, it had some serious disadvantages that ultimately caused its use to dwindle and decline. Hazardous chemicals, including piperidine and hydrazine were needed for the procedure, which put the researchers' safety at risk. The method was both labour and complexity-intensive, requiring exact execution of the chemical processes as well as cautious management of radioactive materials. Additionally, scaling the approach for longer sequences was challenging due to the high quantities of pure DNA required [13]. Early DNA sequencing was greatly aided by Maxam-Gilbert sequencing. It facilitated the acquisition of dependable sequence data and played a pivotal role in the sequencing of the first viral genome, bacteriophage φX174, and the first complete genes, in the lac operon of *Escherichia coli*. These successes paved the way for more sophisticated sequencing technologies, and the main idea driving this was the possibility of using DNA sequencing to uncover, read, and decipher the genetic code! Maxam-Gilbert sequencing is still regarded as a key approach in the history of DNA sequencing, despite being subsequently surpassed by the Sanger method. It substantially contributed to our knowledge of the structure and function of DNA, opening the door for rapid developments in genomics that have subsequently transformed biological science [14].

3.2 Sanger sequencing

Sanger sequencing was one of the first wildly successful sequencing technology developed and is one of the most significant advances in molecular biology. This technique, which was created in 1977 by Frederick Sanger and his associates, served as a foundation for DNA sequencing as a whole and is still a vital component of genetic research today. Our knowledge of genetics was absolutely transformed by Sanger sequencing's capacity to precisely identify the sequence of DNA molecules, which made it possible to conduct in-depth analyses of genes, genomes, and their variants. This technique cemented its place in scientific history pivotally supporting scientific accomplishments including the Human Genome Project and the first whole genome sequencing. Before Sanger sequencing, it was difficult to pinpoint the exact nucleotide order in DNA. Even after Watson and Crick deciphered the DNA structure in 1953, the problem of translating the genetic code remained unresolved. The creation of a trustworthy technique for sequencing DNA created new avenues for investigation, enabling researchers to follow evolutionary trends, investigate the genetic roots of disorders, and even create novel biotechnologies. Sanger's technique, which is known for its simplicity and accuracy, has been used to sequence innumerable genes and genomes over the past 40 years, making it the gold standard for DNA sequencing [5, 11]. DNA synthesis involves the selective integration of chain-terminating dideoxynucleotides (ddNTPs), which forms the basis of the Sanger sequencing method. Sanger sequencing is also referred to as the chain-termination method. Using a DNA polymerase, which creates a complementary strand by progressively adding nucleotides, this technique is based on the enzymatic replication of a single-stranded DNA template. Sanger's approach is notable for its innovative usage of dideoxynucleotides, which are modified nucleotides that do not have the 3′-hydroxyl group needed to create a phosphodiester link with the following nucleotide. Dideoxynucleotides arrest and terminate chain growth as soon as they are incorporated into the growing chain and this stops further elongation of the DNA strand whenever they are integrated [5, 13].

The first step in the Sanger sequencing procedure is to prepare a DNA template, usually a single-stranded DNA fragment representing the target area to be sequenced. DNA synthesis gets underway when a short primer anneals to the template. A mixture of regular deoxynucleotides (dNTPs) and a tiny amount of each of the four dideoxynucleotides (ddATP, ddTTP, ddCTP, and ddGTP), each labeled with a different fluorescent dye, are included in the reaction mixture together with the DNA template, primer, and DNA polymerase [5]. The DNA polymerase extends the primer by adding nucleotides complementary to the template strand during the synthesis step. Periodically, the polymerase mixes in a dideoxynucleotide rather than a regular deoxynucleotide when it adds nucleotides. This results in the termination of that specific strand's synthesis and the accumulation of different-length DNA fragments, all of which have a labeled dideoxynucleotide at the end. The mixture of terminated fragments is separated according to size using capillary electrophoresis, which establishes the DNA sequence [5]. A laser activates the fluorescent labels on the fragments as they travel through the capillary, and a detector identifies the emitted wavelength of light. Each fragment's terminal nucleotide can be distinguished from the others by looking at the color (wavelength) of the fluorescence, as each dideoxynucleotide is labeled with a distinct dye. Subsequently, the sequence is pieced back together by reading the fluorescence pattern from the shortest to the longest segment, which matches the template DNA's sequence [5, 13]. This unravels the original sequence of DNA nucleotides, how they were arranged in the DNA fragment.

With error rates as low as 1 in 10,000 nucleotides, the Sanger method is exceptionally accurate and may be used for a wide range of tasks, including sequencing small genomes and tiny genes. Sanger sequencing is still an essential tool in any molecular biologist's toolkit. Many labs still rely on it for its dependability, even with the advent of more recent high-throughput sequencing methods. This is especially true for projects requiring high accuracy or sequencing relatively short DNA segments [13]. Sanger sequencing has had a significant impact on the fields of genetics and molecular biology. It was the first technique to reliably determine the sequence of DNA, which led to important scientific discoveries and new study directions. In 1977, one of the most significant and pioneering accomplishments with Sanger sequencing was the genome sequencing of the bacteriophage φX174. This was the first complete genome to be sequenced, proving that Sanger's method could sequence an organism's DNA from start to finish. This achievement marked a turning point in genetics and paved the way for larger-scale sequencing initiatives [5, 11].

One of the most ambitious scientific projects of the twentieth century, the Human Genome Project (HGP), relied heavily on Sanger sequencing. The Human Genome Project (HGP) was started in 1990 and finished in 2003 to sequence the entire genome, which has over 3 billion base pairs. Sanger sequencing was a major

component of the endeavor since it produced the high-quality sequence data required to assemble the human genome. Our understanding of human genetics and disease has revolutionized since the successful completion of the Human Genome Project (HGP), which produced a complete reference human genome [11]. Sanger sequencing has several uses beyond ambitious initiatives like the Human Genome Project. It has been extensively employed in medical genetics to find mutations linked to hereditary illnesses. Sanger sequencing, for instance, has been helpful in the diagnosis of diseases like cystic fibrosis, Huntington's disease, and certain types of cancer. Physicians can provide more precise diagnoses, accurate prognostic information, and focused medicines particular to each patient's genetic profile by detecting specific mutations in disease-causing genes. Besides its uses in medicine, Sanger sequencing has greatly influenced evolutionary biology. Scientists have improved our understanding of the tree of life by reconstructing the evolutionary links between animals by comparing DNA sequences from various species. Sanger sequencing has proven to be an invaluable instrument for identifying and describing diseases in microbiology [13–16].
In the study of microbiology, Sanger sequencing has proven to be an invaluable instrument for identifying and describing diseases. The study of infectious diseases has been completely transformed by the capacity to sequence the DNA of microbes, which has sped up the discovery of infections and facilitated the creation of diagnostic tools and therapies. For instance, in 2003, Sanger sequencing was utilized to determine the cause of severe acute respiratory syndrome (SARS), which improved our knowledge of the virus and guided public health measures. By analyzing their sequences, Sanger sequencing confirmed that the intended alterations have been successfully inserted into genetically modified organisms. Gene therapy, genetically engineered crops, and biotechnological uses have all benefited significantly. Sanger

Parameter	Maxam-Gilbert sequencing	Sanger sequencing
Principle	Chemical modification and cleavage of DNA at specific bases	Chain termination using ddNTPs
Read length	~100–500 bp	~500–1000 bp
Accuracy	Moderate (~90–95%)	High (99%)
Error profile	Sensitive to chemical degradation and sequencing errors	Low error rate
Throughput	Low, manual process, slow and labour-intensive	Low to moderate
Speed	Slow complex process requiring hazardous chemicals	Moderate
Cost	High (labour-intensive)	Moderate to high, cheaper with automation
Pros	Prior DNA amplification not required, directly sequences the chemically modified bases	High accuracy, reliable, widely used in HGP and other major projects
Cons	Complex, laborious, lower read lengths, hazardous chemicals	Low throughput, expensive for large projects
Best applications	Early DNA sequencing work, led to development of first-generation sequencing methods, phased out and rarely used today	Ideal for low-throughput projects, great for sequencing PCR products, plasmids

Table 1.
Comparative analysis of Maxam-Gilbert and Sanger sequencing (first generation sequencing platforms).

sequencing is an indispensable tool for quality control in various domains because of its accuracy and dependability [13–16].

To sum up, Sanger sequencing has dramatically influenced the fields of genetics and molecular biology. It was crucial to the Human Genome Project, made possible the first whole genome sequencing, and has been extensively utilized in the fields of medical genetics, evolutionary biology, microbiology, and biotechnology. The technique is still in use in labs worldwide despite the introduction of more advanced technology. This is because of its precision and dependability, making it the gold standard in DNA sequencing. Sanger sequencing has left a significant impact since it helped establish the basis for our present knowledge of the genetic code and continues to influence scientific inquiry and practice across many domains [13–16]. A comparative analysis between Maxam-Gilbert and Sanger sequencing is shown in **Table 1**.

4. Second-generation sequencing

High-throughput sequencing of DNA and RNA was made possible by second-generation sequencing, also referred to as next-generation sequencing (NGS), which completely transformed the field of genomics. In stark contrast to first-generation sequencing, NGS methods enable simultaneous sequencing of millions of DNA fragments. This makes it possible to swiftly and cost-effectively sequence complete genomes. The development of NGS has revolutionized several fields of biological research, including the study of biodiversity and evolutionary links as well as the genetic foundation of illness. NGS's potency lies in its capacity to produce enormous volumes of data, which enable previously unheard-of-scale analyses of genetic variation, gene expression, and epigenetic changes [17, 18]. The capacity for massively parallel sequencing is a defining feature of NGS technologies. NGS technology can sequence millions of DNA fragments in parallel rather than one fragment at a time, significantly boosting throughput and lowering the cost per base. Most NGS technologies work on amplifying DNA fragments, attaching them to beads or a solid surface, and then sequencing them by ligation or synthesis, depending on the platform. Once the sequence data is generated, the pieces are put together into a coherent sequence, and the biological meaning of the findings is deduced by analysis utilizing bioinformatics tools [17–20].

In 2005, the first commercial platform, the 454 Pyrosequencer was released, and the development of NGS got started. Other platforms quickly followed, such as sequencing by ligation (SBL)-using SOLiD system, semiconductor-based Ion Torrent and Illumina's sequencing by synthesis (SBS) technology. These platforms all offered distinct benefits and helped accelerate the development of NGS. Numerous genomic applications now choose NGS because of the quick advancements in accuracy, read length, and cost-efficiency brought about by the competition among NGS technologies [18, 19, 21]. It is not possible to overstate the influence that NGS has had on biological research. Large-scale initiatives like the Earth Microbiome Project, which aims to characterize microbial variety throughout the globe, and the 1000 Genomes Project, which cataloged human genetic variation, have been made possible by it. Personalized genomics—a medical field where a patient's genetic information can inform clinical decisions on everything from the diagnosis of genetic abnormalities to the customization of cancer treatments—was made possible by NGS. The discovery of new genes, the investigation of complex traits, and the study of ancient DNA have all accelerated due to the rapid and affordable sequencing of complete genomes [1, 19–21].

4.1 Pyrosequencing (454 sequencing)

Introduced in 2005, the 454 Genome Sequencer was the first NGS platform to be sold commercially. The technique of Pyrosequencing, which is based on the detection of pyrophosphate release during DNA synthesis was introduced by this platform. This technology was created by 454 Life Sciences, and had substantially better throughput and quicker turnaround times than standard Sanger sequencing [1, 17]. Pyrosequencing synthesizes a complementary DNA strand by adding nucleotides (A, T, C, or G) one after the other from a single-stranded template. Every nucleotide that joins the expanding DNA strand releases a pyrophosphate molecule. After the enzyme sulfurylase transforms this pyrophosphate into ATP, it catalyzes a process that emits light. The sequence may be deduced from the light signal since the intensity of the light released is proportionate to the number of nucleotides included. DNA fragments on beads were amplified using emulsion PCR (emPCR) on the 454 pyrosequencing platform and then put into picotiter plates for sequencing [17].

Together with the enzymes needed for pyrosequencing, a single DNA-bound bead was present in each well of the picotiter plate. The light emitted was recorded as nucleotides were added one after the other across the plate, resulting in a readout of the sequence for each DNA fragment [17]. One of its main advantages was the comparatively long read lengths of 454 pyrosequencing, which were distinctively longer than those generated by competing NGS systems at the time. Because of this, the 454 platform proved to be especially helpful for de novo genome assembly and sequencing genomic sections containing intricate repeats. But the platform also has certain drawbacks, such as a greater mistake rate in homopolymeric regions—stretches of nucleotides that are identical to one another—caused by challenges in precisely calculating the quantity of nucleotides inserted. Notwithstanding these drawbacks, 454 pyrosequencing greatly influenced the genomics community. It was utilized in numerous ground-breaking investigations, such as the metagenomic analysis of microbial populations and the NGS sequencing of the first human genome. However, 454 pyrosequencing's use declined as competing NGS methods advanced and became more affordable, and the platform was eventually shut down in 2016 [1, 17].

4.2 Illumina sequencing

Sequencing by synthesis, or Illumina sequencing, is the most commonly used NGS technology when writing this chapter (Sep 2024). Since its introduction in 2006, Illumina's platform—which boasts excellent accuracy, scalability, and cost-effectiveness—has quickly emerged as the industry standard for NGS technology. The method involves synthesizing a complementary DNA strand from a template by introducing fluorescently labeled nucleotides one at a time. Because each nucleotide has a distinct fluorescent dye labeled to it, the sequence is established by observing the fluorescence color following each inclusion [18, 19]. The DNA sample is first broken up using an enzymatic process "tagmentation" process. This is a random process that creates millions of fragments. The ends of the fragments are fitted with adapters (defined short DNA sequences) to start the process. These adapters are utilized to secure the fragments to a solid surface—usually a glass flow cell—so that bridge amplification can be used to enhance their signal. This amplification creates clusters of identical DNA fragments are produced by this amplification process, and they are sequenced simultaneously. Fluorescently labeled nucleotides are introduced to the flow cell during the sequencing operation. A laser excites the fluorescent label when each nucleotide is added to the expanding DNA strand, and a camera records the emitted light. The color of the fluorescence indicates the identity of the nucleotide. The fluorescent label is chemically separated from the nucleotide at the end of each cycle, enabling the addition of the subsequent nucleotide. For every sequencing cycle, this procedure is repeated, producing a sequence read for every DNA fragment [19].

Illumina sequencing is highly accurate, with error rates often less than 0.1%, which is one of its main advantages. The platform can produce billions of readings in a single run and provides extremely high throughput. Because of this, it is perfect for a variety of uses, including targeted sequencing of particular genomic areas and whole-genome sequencing. Furthermore, paired-end sequencing—in which a DNA fragment's two ends are sequenced—is supported by Illumina sequencing. This technique increases the precision of genome assemblies by revealing more information about the sequence [19]. Numerous extensive genomic initiatives, such as the Human Microbiome Project, the Cancer Genome Atlas, and the 1000 Genomes Project, have made use of Illumina sequencing. Additionally, it has been widely used in clinical settings for a variety of purposes, including the diagnosis of genetic diseases, cancer genomics, and non-invasive prenatal diagnostics. Numerous academics and physicians have chosen the platform because of its affordability, scalability, and adaptability [18, 19].

4.3 SOLiD sequencing

Another significant second-generation sequencing technique was SOLiD (Sequencing by Oligonucleotide Ligation and Detection) sequencing, which was created by Applied Biosystems (now a division of Thermo Fisher Scientific) [22]. SOLiD sequencing was first introduced in 2007 and is centred on the idea of sequencing by ligation, which involves ligating short, fluorescently labeled oligonucleotides in a sequence-specific way to a developing DNA strand [22]. The SOLiD platform employs a special two-base encoding scheme in which a pair of colors is used to identify each nucleotide. As a result of the sequencing process reading each nucleotide twice, this technology has built-in mistake correction [22, 23]. The DNA sample is broken up and adapters are attached to the ends of the broken pieces to start the sequencing process. After immobilizing the fragments on a solid surface—usually beads—emulsion PCR is used to amplify them [22, 23].

A pool of degenerate oligonucleotides that have been fluorescently labeled is added throughout the sequencing operation. These oligonucleotides have a cleavage site between the first and second bases and are complementary to the template DNA. The fluorescent signal is observed when the oligonucleotide that matches the sequence at the first location is ligated to the developing DNA strand. The oligonucleotide is cleaved at the cleavage site, and the procedure is then repeated for the subsequent position [23]. SOLiD sequencing's two-base encoding scheme offers excellent accuracy, with error rates usually less than 0.1%. The platform was less useful for several applications, like de novo genome assembly, because its read lengths were sometimes shorter than those of other NGS methods. Even while SOLiD was accurate, its popularity declined due to its complicated methodology and the rise of more user-friendly platforms like Illumina. Though it was subsequently terminated, the SOLiD platform was crucial to the early advancement of NGS technologies [22–24].

4.4 Ion Torrent sequencing

Ion Torrent is an NGS technology that uses a semiconductor-based detection system. It was introduced by Life Technologies in 2010, and marked a significant departure from the more common optical based NGS methods. At the core of its innovation lies a semiconductor [1, 18, 19]. This ingenious approach is based on detecting Hydrogen ions released during DNA polymerization, allowing the sequence to directly measure nucleotide incorporation without the need for fluorescent labels. The platform uses a microchip containing millions of individual wells, with each well housing a single DNA template attached to a bead. As nucleotides sequentially flow across the chip, DNA polymerase adds nucleotides to the growing strand. This releases a Hydrogen Ion, causing a slight pH change, which is detected by sensors placed on the chip. The signal corresponds to the specific nucleotide incorporated, allowing the sequence to be determined in real time. The main advantage of Ion Torrent Technology is its scalability and speed, with rapid sequencing turnaround times, often assembling data in a matter of hours. The platform is highly flexible with the ability to sequence small genomes and PCR products. The platform does have limitations, particularly with homopolymer regions (long stretches of the same nucleotide). These regions can cause difficulties in accurate base calling, due to the inherent nature of the pH-based detection system, leading to sequencing errors in the form of insertions or deletions. The Ion Torrent platform has improved with time, offering various models to meet different sequencing needs. Early models were designed for small-scale projects, while later models offered increased throughput and scalability for larger Whole Genome Sequencing based projects. Ion Torrent is widely used in a variety of applications, which include microbial sequencing, clinical diagnostics and cancer research for its distinct advantages of speed, simplicity and affordability. It continues to be a significant player in the field of next generation sequencing [1, 18, 19].

4.5 Comparative analysis of second-generation sequencing technologies

The field of genomics has benefited from the distinct advantages and disadvantages that second-generation sequencing technologies, such as 454 pyrosequencing, Illumina sequencing, Ion Torrent and SOLiD sequencing, have each brought to the table. These technologies were adopted and applied in different research contexts based on trade-offs between accuracy, read length, throughput, and cost, which are highlighted in a comparative analysis of these technologies [1, 18, 19, 22, 24]. With error rates usually less than 0.1%, Illumina sequencing proved to be the most accurate of the three methods. Because of this, it was the go-to option for tasks requiring precision, like extensive population genomics research and clinical diagnostics. Because SOLiD sequencing used a two-base encoding technique with integrated error correction, it also delivered great accuracy. Nevertheless, compared to Illumina, its application was occasionally limited by the shorter read lengths and more complicated procedure. In contrast, 454 pyrosequencing encountered difficulties in precisely sequencing homopolymeric regions, resulting in greater error rates in certain locations, while being a pioneer in its high throughput and longer read lengths. On the contrary, Ion Torrent's lack of optical detection allows for rapid sequencing, making it ideal for clinical applications where turnaround time is crucial. The system is less expensive making it an attractive option for smaller laboratories. Like 454, Ion Torrent also struggles with long stretches of homopolymer regions [1, 18, 19, 22].

Another important consideration when selecting a sequencing method is read length. A prominent feature of 454 pyrosequencing was its comparatively high read lengths, which frequently exceeded 400 base pairs. This proved useful for sequencing repetitive sequence areas and building complicated genomes. Because of this, 454 was appropriate for long read applications such as de novo genome assembly. Ion

Torrent produced reads in the range of 200–400 bp with average read length shorter than that of 454 but still longer than the read lengths produced by Illumina. At first, Illumina sequencing only provided shorter read lengths, usually between 75 and 150 base pairs. But thanks to improvements in Illumina technology, read lengths can now reach over 300 base pairs, making them more useful for a wider range of tasks, such as transcriptome analysis and genome assembly. SOLiD sequencing generally produced shorter reads, typically between 35 and 50 base pairs, which limited its usefulness for de novo assembly but made it useful for applications needing high accuracy and error correction [1, 18, 19, 22]. A sequencing technology's throughput is the volume of data it can produce in a single run, and for large-scale projects, cost-effectiveness becomes a crucial factor. One notable feature of Illumina sequencing is its tremendous throughput, as it can produce billions of reads in a single sequencing run. Because of its high throughput and falling sequencing base cost, Illumina is now the most extensively used NGS platform. High throughput was also provided by 454 pyrosequencing, but because it cost more per base than Illumina, it lost favor when Illumina technology became more affordable. Although Ion Torrent was fast, cost effective and scalable, its high homopolymer error rate caused it to gradually lose ground. In comparison to Illumina, SOLiD sequencing offered a moderate throughput at a usually higher cost per base. SOLiD was generally more expensive per base than Illumina, despite having higher precision. For some applications, however, its shorter read lengths and higher cost rendered it less appealing than the more affordable and scalable Illumina platform [1, 18, 19, 22].

The adoption of a sequencing platform can be greatly influenced by its workflow and user-friendliness. The workflow of Illumina sequencing is renowned for being user-friendly, featuring simple procedures for data processing, sequencing, and library preparation. Illumina technology's high throughput and precision, in addition to its simplicity, have made it popular in both clinical and research contexts. Although novel, 454 pyrosequencing necessitated more intricate library preparation and data processing, which may be difficult for certain users. Similarly, Ion Torrent had notable limitations, especially in the context of read accuracy in sequencing homopolymer regions. The system measures amount of pH changes and this aspect became problematic during sequencing repeat regions in complex genomes. Another limitation was the reads it produced in the range of 200–400 bp, which were shorter than the reads produced by comparable platforms. This eventually limited its utility in applications like resolving structural variants and assembly of complex genomes. The oligonucleotide ligation and fluorescent detection processes in the SOLiD sequencing technique were complicated as well, requiring several steps. This intricacy led to the platform's demise in favor of more approachable technologies like Illumina, combined with its lower throughput and greater cost [1, 18, 19, 22]. The particular needs of the application are a major factor in selecting the right sequencing technology. Whole-genome sequencing, RNA sequencing, and epigenomics are just a few of the many applications that have made Illumina sequencing the industry standard thanks to its high accuracy, long read lengths, and affordability. Large-scale initiatives like the Cancer Genome Atlas and the 1000 Genomes Project have benefited greatly from it. Although 454 pyrosequencing's greater cost and error rate in homopolymer regions limited its utilization over time, its longer read lengths made it especially useful for de novo genome assembly and metagenomics. With its distinct two-base encoding approach and high accuracy, SOLiD sequencing established a niche market for mistake repair projects; however, due to its greater cost and shorter read lengths, it was less competitive than Illumina [1, 18].

Parameter	Roche 454	Illumina	SOLiD	Ion Torrent
Principle	Pyrosequencing (light detection via pyrophosphate release)	Sequencing by synthesis	Sequencing by ligation (color-space detection)	Semiconductor based detection of pH change
Read length	Up to 700 bp	100–300 bp (paired end up to 600 bp)	50–75 bp	200–400 bp
Accuracy	Moderate, issues with homopolymers	Very high (~99%)	Very high (~99%)	Moderate, issues with homopolymers
Error profile	Homopolymer errors (insertions and deletions)	Low error rate, high accuracy	Color-space encoding prevents substitution errors	Homopolymer errors (insertions and deletions)
Throughput	Moderate (~1 Gb/run	Very high (up to 6 Tb/run with NovaSeq)	High (up to ~300 GB/run)	Moderate (up to ~1 GB/run)
Speed	Moderate	Slower but higher throughput	Slow (long run times)	Fast (results within hours
Cost	High (expensive reagents and instrument)	Moderate to high (depending on the scale)	High (expensive reagents and instrument)	Low to moderate (affordable equipment)
Pros	Longer read lengths, good for de novo sequencing	High accuracy, high throughput, broad application range	High accuracy with built in error correction	Fast, affordable and scalable
Cons	Homopolymer errors, high cost, discontinued	Expensive for long reads, limited by short reads for structural variants	Short reads, slow throughput, complex workflow	Short reads, homopolymer errors
Best applications	Metagenomics, de novo assembly	Whole genome sequencing, RNA and exome sequencing	SNP detection, targeted sequencing (small)	Microbial sequencing, clinical diagnostics

Table 2.
Comparative analysis of various second generation (next-generation sequencing) platforms.

In conclusion, despite the fact that every second-generation sequencing technology has particular benefits and drawbacks, Illumina sequencing became the most widely used platform because of its excellent accuracy, long read lengths, high throughput, and affordability. The growth and advancement of NGS technologies have significantly increased the potential of genomic research by giving scientists strong instruments to investigate the genetic foundation of life and deepen our knowledge of biology. A comparative analysis of second-generation sequencing or NGS platforms is shown in **Table 2**.

5. Third-generation sequencing or single molecule sequencing (SMS)

Single-molecule sequencing (SMS) technologies, commonly referred to as third-generation sequencing technologies, are a major improvement over second-generation sequencing technique. Third-generation sequencing, which analyses individual DNA molecules directly as opposed to previous sequencing techniques that rely on the amplification of DNA fragments, has a number of important benefits, including the capacity to capture epigenetic alterations, longer read lengths, and less complexity. By enabling more precise genome assemblies, improved characterization of intricate genomic areas, and better understanding of structural variants and epigenetic landscapes, these technologies have completely transformed the field of genomics [25]. Some of the shortcomings of second-generation sequencing techniques, such as the challenges of sequencing lengthy repetitive areas and the biases generated during amplification, are addressed by single-molecule sequencing technology. These technologies enable the investigation of biological processes that were previously difficult to probe and offer a more comprehensive perspective of the genome by directly sequencing DNA molecules. The two most well-known third-generation sequencing platforms are Oxford Nanopore Technologies and Pacific Biosciences (PacBio). Both systems provide distinctive methods for single-molecule sequencing and have made major contributions to the fields of personalized medicine and genomics [25, 26].

5.1 PacBio

PacBio, also known as Pacific Biosciences, developed Single Molecule, Real-Time (SMRT) sequencing technology, which revolutionized single-molecule sequencing. Since its initial introduction in 2009, PacBio's platform has emerged as a prominent technology in the field of third-generation sequencing, owing to its capacity to produce lengthy reads and furnish extensive genomic data. Real-time sequencing of individual DNA molecules is the foundation of PacBio's SMRT sequencing technology [27]. The device makes use of a novel technique known as zero-mode waveguide (ZMW) to make real-time monitoring of DNA polymerase activity possible. DNA molecules are adhered to a surface inside a ZMW—a tiny, nanoscale chamber—during SMRT sequencing in order to detect the fluorescence released after nucleotide incorporation [27]. DNA fragments are affixed to SMRT cells to create a DNA library, which is the first step in the sequencing process. Thousands of ZMWs, each containing a single DNA molecule and the enzyme DNA polymerase, are found inside the SMRT cell. Fluorescently labeled nucleotides are added one by one while the DNA polymerase creates a complementary strand. A camera tracks the fluorescence that each nucleotide emits, giving real-time information about the DNA sequence [27, 28]. With PacBio's technology, read durations can be remarkably extended, frequently surpassing 10,000 base pairs and occasionally even reaching 30,000 base pairs. The capacity to read long sequences is very helpful for de novo genome assembly, as longer reads can bridge repeated sections and reconcile intricate structural variations. Furthermore, real-time sequencing provides extensive information about the sequence as it is read and enables ongoing monitoring of DNA synthesis [28].

The capacity of PacBio's SMRT sequencing technique to produce lengthy reads is one of its main benefits. Accurately assembling complicated genomes, particularly ones with structural changes or repetitive areas, requires long reads. PacBio is now a useful tool for transcriptome analysis, epigenomics, and genome assembly, among other applications. Additionally, PacBio's method offers high base calling accuracy, with error rates in raw reads typically ranging from 10% to 15%. These error rates can be further reduced by using consensus sequencing and error-correction algorithms. PacBio's SMRT sequencing does, however, have several drawbacks [27]. Large-scale initiatives may find the technology prohibitive because of its comparatively greater cost per base when compared to second-generation sequencing techniques. Furthermore, PacBio's platform has a lower throughput than some second-generation sequencing technologies, which would make it less useful for some high-throughput applications. PacBio's technology has significantly advanced genomics despite these drawbacks. It has been applied to a number of research endeavors, such as the characterization of microbial communities, the study of structural changes in cancer genomes, and the sequencing of complex plant and animal genomes. PacBio's platform's long-read capabilities has also made it possible to find new genomic features and shed light on the genetics of complicated features [27, 28].

5.2 Oxford Nanopore Technologies

Using nanopore-based technology, Oxford Nanopore Technologies (ONT) has created a novel method for single-molecule sequencing. The MinION sequencer, the company's flagship device, was first released in 2014 and has subsequently gained popularity as a third-generation sequencing platform. The technology of ONT is distinguished by its scalability, mobility, and capacity to produce lengthy readings [6]. The basis of ONT's sequencing technology is the measurement of ionic current variations as DNA molecules are pulled through a protein pore of nanometer measurements—hence the name nanopore. The sequence can be established by measuring the alterations in ionic current that result from the translocation of DNA molecules via a nanopore. The first step in the sequencing procedure is to create a DNA library, in which DNA fragments are joined to a motor protein to help move the DNA through the nanopore. The motor protein regulates the DNA's flow through the nanopore, resulting in an accurate and productive sequencing process. A distinctive signal that is unique to each nucleotide is produced when the DNA obstructs the ions' passage through the nanopore. The DNA sequence is read by detecting and examining this signal. ONT's method produces very long reads, frequently more than an astonishing 100,000 base pairs, which is very useful for studying structural changes and de novo genome assembly [6, 29].

The capacity of ONT's sequencing technique to produce extraordinarily lengthy reads remains one of its main advantages. Sequencing genomes containing repetitive areas, long-range haplotypes, and complicated structural variants benefits greatly from the long-read capacity. ONT's technology is also incredibly scalable and portable; and can provide real-time sequencing data and is compact enough to carry in a laptop bag! The comparatively low cost per base of ONT's technology is another benefit that makes it a desirable choice for high-throughput data applications and large-scale sequencing operations. Real-time sequencing facilitates quick data collection and analysis, which is helpful for applications like infectious disease surveillance and field-based genomics [6]. ONT's sequencing technology is not without its shortcomings. Compared to other sequencing technologies, the accuracy of base calling in raw reads is poorer, with error rates usually ranging from 5% to 10%. Applications needing a high degree of precision, such clinical diagnostics and variation detection, may find this difficult. Furthermore, the technology is still developing, and throughput, accuracy, and read duration are all constantly improving. Notwithstanding these difficulties, ONT's technology has significantly advanced genomics. Numerous research applications have made use of it, such as the analysis of complicated plant genomes, the study of cancer genomics, and the sequencing of microbial genomes. ONT's platforms' scalability and portability have also created new avenues for genomic research in a variety of environments [6, 29].

5.3 Comparative analysis of third generation sequencing technologies

The comparison of PacBio and Oxford Nanopore Technologies highlights the strengths and limitations of each platform in the context of single-molecule sequencing. Both technologies offer unique advantages, and their choice depends on the specific requirements of the research or application. Long readings are a strength of both Oxford Nanopore Technologies and PacBio. While ONT's technique can yield reads surpassing 100,000 base pairs, PacBio's SMRT sequencing method normally delivers reads between 10,000 and 30,000 base pairs. Long read production is useful for de novo genome assembly, structural variation analysis, and repetitive region characterization. For some applications, ONT's technology can be helpful due to its distinct advantage in read length [6, 27–29]. Comparing ONT's nanopore-based technology to PacBio's SMRT sequencing technique, the latter often provides base calling accuracy that is higher. The error rate in raw reads for PacBio is usually between 10% and 15%, while error-correction methods can make the rate lower. In contrast, ONT's technology has a higher raw error rate, usually between 5% and 10%. ONT's technology's lesser accuracy may be a hindrance for high precision applications like clinical diagnostics and variant detection [6, 27–29]. Large-scale projects may want to take into account that PacBio's technology has a greater cost per base than ONT's. ONT's MinION sequencer is a desirable choice for high-throughput applications because it is reasonably priced and has a low cost per base. While ONT's technology offers high throughput and real-time sequencing capabilities, PacBio's throughput is less than that of other second-generation sequencing methods [6, 27–29]. The MinION

Parameter	PacBio (SMRT)	Oxford nanopore
Principle	Single-molecule Real Time sequencing using zero-mode waveguides	Single-molecule via nanopore-based detection of electrical charges
Read length	10,000–100,000 bp	Ultra-long reads (up to 2 Mb, average read length of 10,000 bp)
Accuracy	HiFi reads >99.9% (high accuracy)	>99.9% (high accuracy)
Error profile	Raw error rate high but improved with HiFi reads	Moderate error rate in homopolymer and repetitive regions, improves with depth of coverage
Throughput	Moderate (~500 Gb per run)	Very high (several Tb on the PromethION)
Speed	Slow due to long reads and real-time platform	Fast, especially with small genomes
Cost	High per run, especially for HiFi Sequencing	Low to moderate (affordable and portable equipment, reagents vary by scale
Pros	Ultra-long reads, high accuracy with HiFi reads,	Ultra-long reads, portable and scalable, flexible for both small and large projects
Cons	High raw-error rate, expensive	Raw reads with moderate accuracy, complex data analysis
Best applications	Structural variant detection, complex genome assembly, full-length RNA sequencing	Long-read sequencing, portable sequencing

Table 3.
Comparative analysis between third-generation sequencing technology PacBio and Oxford Nanopore.

sequencer is tiny and able to provide real-time sequencing data, and ONT's technology is very portable and scalable. Applications in rural areas and field-based genomics benefit from this portability. Although PacBio's technology is quite accurate and can produce lengthy scans, it requires a more substantial infrastructure and has a wider footprint [6, 27–29]. A comparative analysis between PacBio and Oxford Nanpore Sequencing Technology is shown in **Table 3**.

6. Beyond the sequence: The lasting impact of DNA sequencing

DNA sequencing has had an enormous far-reaching impact on a number of disciplines, including biology, medicine, agriculture, and even forensic research. Although the 1950s saw the discovery of the DNA double helix, which provided the groundwork for understanding genetic inheritance, the development of DNA sequencing revolutionized our understanding of and interactions with the life-code. DNA sequencing has changed over the past few decades from a labour-intensive, complex process to a highly automated, accessible technology that has sparked tremendous breakthroughs in a wide range of fields. It is a transformational technology that has completely altered the way we think about biological research and clinical practice by providing previously unattainable insights [11]. DNA sequencing has shown to be a very useful technique for the diagnosis and treatment of genetic problems. Many diseases having genetic roots could only be studied indirectly or with inadequate approaches prior to the advent of sequencing tools. Today it is possible to pinpoint the precise mutations that cause diseases like muscular dystrophy, cystic fibrosis, and different types of cancer. As a result, tailored medicines have been developed, especially in the field of gene therapy, which treats illness fundamentally by replacing or correcting defective genes. The development of personalized medicine also depends heavily on DNA sequencing. Doctors can now offer focused approach to healthcare by customizing medications to each patient's unique genetic profile. DNA sequencing makes it possible to treat cancer more successfully than with traditional chemotherapy by allowing them to pinpoint the genetic alterations present in a tumor and develop therapies that particularly target those changes [25].

Our knowledge of human evolution and evolutionary biology in general has been revolutionized by DNA sequencing. By analyzing genomes from various species, researchers can trace evolutionary relationships and shed light on the genetic alterations that lead to speciation and adaptation thanks to sequencing technologies. We have now been able to piece together the evolutionary history of life, producing a detailed picture of the millions of years that species have undergone. In order to better understand the genetic similarities and differences between current humans and our evolutionary ancestors, like Neanderthals, we have been able to research extinct animals, one of the most remarkable examples of which is the sequencing of ancient DNA [30]. These findings indicate that early Homo sapiens and Neanderthals interbred, leaving a genetic legacy that is still present in many contemporary human populations. The development of DNA sequencing has also greatly benefited the agriculture sector. Crop resistance to disease, pests, and environmental stresses is critical as the world's population rises and climate change creates new obstacles for food production. Scientists can now detect genetic features in plants that can be used to increase crop yields and quality thanks to DNA sequencing. Genetically modified organisms (GMOs) that are more resilient and more adaptable to shifting environmental conditions are a direct result of this. These developments are especially critical in areas where food insecurity is a significant issue. In these regions, DNA sequencing is assisting farmers in growing more resilient crops that can endure the strains of severe weather, drought, and other climate-related issues [31].

DNA sequencing has become an essential component of contemporary criminal investigations in the field of forensic science. Tiny amounts of genetic evidence from crime scenes can now be analyzed, which has totally changed the way crimes are investigated by improving the accuracy of suspect identification and clearing falsely condemned people. Law enforcement organizations can now more easily identify and apprehend criminals by comparing DNA samples from crime scenes to known offenders thanks to the development of DNA databases. Furthermore, forensic scientists can now analyze even old or deteriorated evidence because to advancements in sequencing technology, which allowed them to solve cold cases that would have remained unsolved otherwise [32]. The integration of machine learning and artificial intelligence (AI) is an intriguing new area in the field of DNA sequencing. With sequencing technologies producing ever-larger amounts of data, artificial intelligence (AI) has the ability to analyze these massive datasets and find patterns that are impossible for humans to find. AI, for example, may be able to assist in the identification of hitherto unidentified genetic variations associated with certain diseases or aid in the understanding of the intricate interactions between genes and environmental variables in the genesis of diseases. This might result in the identification of novel therapeutic targets and more individualized patient care regimens [33].

In the long run, DNA sequencing promises more benefits as costs come down and new technologies are created. Single-cell sequencing is one of the most promising areas of innovation because it enables researchers to examine the genetic composition of individual cells. This is especially crucial in domains such as oncology, where a single tumor may comprise a heterogeneous population of cells with various genetic abnormalities. Through the sequencing of these individual cells, scientists may be able to devise more potent treatments by better understanding how tumors change over time and become resistant to them. In developmental biology, single-cell sequencing is especially useful because it sheds light on how different cells within an organism specialize into distinct tissues after developing from a single fertilized egg [34]. DNA sequencing will continue to influence science and medicine in ways that we have only just begun to fully understand. It has numerous uses and the potential to significantly advance human health, food security, and our comprehension of life. Sequencing will become an even more crucial tool in clinical and scientific contexts as it gets cheaper, faster, and more precise. Advancements in technology in the future, along with the integration of Artificial Intelligence for data analysis and creation of even more cost-effective and rapid sequencing methods, hold the potential to further enhance our comprehension of Genomics. This will lead to new discoveries and advances in a wide range of disciplines.

From the uncovering of the DNA double helix to the advancement of advanced next-generation sequencing technologies, the history of DNA sequencing demonstrates human creativity and cooperation, as each new breakthrough builds upon earlier discoveries, resulting in powerful tools that are continually unveiling the mysteries of the genome. In summary, DNA sequencing has already profoundly changed our understanding of biology, and as the technology develops further, it will surely be central to the next round of scientific discoveries and medical advances.

Acknowledgements

The author acknowledges the use of Grammarly for language polishing of the manuscript.

Author details

Amit Mathews
Canadian Food Inspection Agency, Scarborough, ON, Canada

*Address all correspondence to: amit.mathews@inspection.gc.ca

References

[1] Mardis ER. Next-generation DNA sequencing methods. Annual Review of Genomics and Human Genetics. 2008;**9**:387-402. DOI: 10.1146/annurev.genom.9.081307.164359

[2] Goodwin S, McPherson JD, McCombie WR. Coming of age: Ten years of next-generation sequencing technologies. Nature Reviews Genetics. 2016;**17**(6):333-351. DOI: 10.1038/nrg.2016.49

[3] Watson JD, Crick FHC. Molecular structure of nucleic acids: A structure for deoxyribose nucleic acid. Nature. 1953;**171**(4356):737-738. DOI: 10.1038/171737a0

[4] Franklin R, Gosling RG. Molecular configuration in sodium thymonucleate. Nature. 1953;**171**(4356):740-741. DOI: 10.1038/171740a0

[5] Sanger F, Nicklen S, Coulson AR. DNA sequencing with chain-terminating inhibitors. Proceedings of the National Academy of Sciences. 1977;**74**(12):5463-5467. DOI: 10.1073/pnas.74.12.5463

[6] Jain M, Olsen HE, Paten B, Akeson M. The Oxford Nanopore MinION: Delivery of nanopore sequencing to the genomics community. Genome Biology. 2016;**17**(1):239. DOI: 10.1186/s13059-016-1103-0

[7] Avery OT, MacLeod CM, McCarty M. Studies on the chemical nature of the substance inducing transformation of pneumococcal types: Induction of transformation by a desoxyribonucleic acid fraction isolated from pneumococcus type III. Journal of Experimental Medicine. 1944;**79**(2):137-158. DOI: 10.1084/jem.79.2.137

[8] Holley RW et al. Structure of a ribonucleic acid. Science. 1965;**147**(3664):1462-1465. DOI: 10.1126/science.147.3664.1462

[9] Wu R, Kaiser AD. Structure and base sequence in the cohesive ends of bacteriophage lambda DNA. Journal of Molecular Biology. 1968;**35**(4):523-537. DOI: 10.1016/S0022-2836(68)80012-9

[10] Maxam AM, Gilbert W. A new method for sequencing DNA. Proceedings of the National Academy of Sciences. 1977;**74**(2):560-564. DOI: 10.1073/pnas.74.2.560

[11] International Human Genome Sequencing Consortium. Finishing the euchromatic sequence of the human genome. Nature. 2004;**431**(7011):931-945. DOI: 10.1038/nature03001

[12] Gilbert W, Maxam A. The nucleotide sequence of the lac operator. Proceedings of the National Academy of Sciences of the United States of America. 1973;**70**(12):3581-3584. DOI: 10.1073/pnas.70.12.3581

[13] Brown TA. The Applications of Gene Cloning and DNA Analysis in Research. Gene Cloning and DNA Analysis. Blackwell Publishing; 2006. pp. 197-219. ISBN-10: 1405111216

[14] Venter JC et al. The sequence of the human genome. Science. 2001;**291**(5507):1304-1351. DOI: 10.1126/science.1058040

[15] Shendure J, Ji H. Next-generation DNA sequencing. Nature Biotechnology. 2008;**26**(10):1135-1145. DOI: 10.1038/nbt1486

[16] Smith LM, Sanders JZ, Kaiser RJ, Hughes P, Dodd C, Connell CR, et al.

Fluorescence detection in automated DNA sequence analysis. Nature. 1986;**321**(6071):674-679. DOI: 10.1038/321674a0

[17] Margulies M, Egholm M, Altman WE, et al. Genome sequencing in microfabricated high-density picoliter reactors. Nature. 2005;**437**(7057):376-380. DOI: 10.1038/nature03959

[18] Metzker ML. Sequencing technologies — The next generation. Nature Reviews Genetics. 2010;**11**(1):31-46. DOI: 10.1038/nrg2626

[19] Quail MA, Smith M, Coupland P, et al. A tale of three next-generation sequencing platforms: Comparison of Ion Torrent, PacBio, and Illumina MiSeq sequencers. BMC Genomics. 2012;**13**:341. DOI: 10.1186/1471-2164-13-341

[20] Bentley DR, Balasubramanian S, Swerdlow HP, et al. Accurate whole human genome sequencing using reversible terminator chemistry. Nature. 2008;**456**(7218):53-59. DOI: 10.1038/nature07517

[21] Rothberg JM, Leamon JH. The development and impact of 454 sequencing. Nature Biotechnology. 2008;**26**(10):1117-1125. DOI: 10.1038/nbt1485

[22] Meyer M, Kircher M. Illumina and SOLiD sequencing—Comparison of sequencing platforms and their applications. Nature Reviews Genetics. 2010;**11**(3):169-179. DOI: 10.1038/nrg2739

[23] Garrido-Cardenas JA et al. DNA sequencing sensors: An overview. sensors. 2017;**17**(3):588-602. DOI: 10.3390/s17030588

[24] Nielsen R, Paul JR, Jorgensen TD. The SOLiD sequencing technology. Nature Methods. 2009;**6**(8):621-624. DOI: 10.1038/nmeth.f.261

[25] Pareek CS et al. Sequencing technologies and genome sequencing. Journal of Applied Genetics. 2011;**52**:413-435. DOI: 10.1007/s13353-011-0057-x

[26] Chaisson MJ, Tesler G. Long read sequencing and the future of genomics. Nature Reviews Genetics. 2012;**13**(11):711-723. DOI: 10.1038/nrg3123

[27] Rhoads A, Au KF. PacBio sequencing and its applications. Genomics, Proteomics & Bioinformatics. 2015;**13**(5):278-289. DOI: 10.1016/j.gpb.2015.08.002

[28] Eid J et al. Real time DNA sequencing from single polymerase molecules. Science. 2009;**323**(5910):133-138. DOI: 10.1126/science.1162986

[29] Loman NJ, Watson M. Successful test launch for nanopore sequencing. Nature Methods. 2015;**12**(4):303-304. DOI: 10.1038/nmeth.3327

[30] Green RE et al. A draft sequence of the Neandertal genome. Science. 2010;**328**(5979):710-722. DOI: 10.1126/science.1188021

[31] Shendure J et al. DNA sequencing at 40: Past, present and future. Nature. 2017;**550**:345-353. DOI: 10.1038/nature24286

[32] Butler JM. DNA databases: Uses and issues. In: Advanced Topics in Forensic DNA Typing Methodology. Elsevier; 2012. pp. 213-270. ISBN-10: 0123745136

[33] Zou J et al. A primer on deep learning in genomics. Nature

Genetics. 2019;**51**:12-18. DOI: 10.1038/s41588-018-0295-5

[34] Macosko EZ et al. Highly parallel genome-wide expression profiling of individual cells using nanoliter droplets. Cell. 2015;**161**(5):1202-1214. DOI: 10.1016/j.cell.2015.05.002

Chapter 2

The First Years of DNA Sequencing

Mevlüt Arslan

Abstract

We were able to sequence DNA molecules a 100 years later, after its discovery. Wu was the first scientist to report DNA sequencing in 1968. In 1970, we learned the first eight-base sequence of lambda DNA from Wu's studies, which were based on synthesis by DNA polymerase to be sequenced DNA. In 1971, a four-base sequence was added also, and we learned a twelve-base sequence in total. In the same year, Englund reported a direct DNA sequencing method that allowed *in vivo* DNA to be sequenced directly without synthesis. After 1971, DNA sequencing studies have developed in two ways: *sequencing by synthesis* and *direct sequencing*. Sanger reported the first DNA sequencing study in 1973, when Wu introduced the primer-extension method to sequence-specific DNA regions. Although Wu's and other scientists' studies were very valuable, the techniques were difficult. Sanger's methods described in 1973 and 1975 were also difficult. Practical methods were highly demanded. In 1977, Sanger introduced a more practical chain termination method, also known as the "dideoxy method," which was related to *sequencing by synthesis*. Direct sequencing of *in vivo* DNA was highly desired. In the rout, Maxam-Gilberts introduced the chemical-cleavage method in 1977. Seven years after the first DNA sequencing, DNA sequencing methods reached a significant milestone.

Keywords: DNA sequencing, first years, Ray Wu, Sanger, Maxam-Gilbert, *direct DNA sequencing*, *sequencing by synthesis*

1. Introduction

Deoxyribonucleic acid (DNA) is the material that determines all biological activities and properties in the cell. Therefore, understating or characterizing DNA molecules is very important in all biological science. Our understanding of DNA properties has mostly been achieved through DNA sequencing studies. From the discovery of the DNA molecule to its sequencing, it required about a 100 years.

The history of DNA-related science was opened by Frederic Miescher in 1868. Miescher studied lymphocytes obtained from pus in fresh surgical bandages. In one study, Miescher isolated a substance that exhibited different properties rather than expected. The most interesting property of this substance was that it was not stained by iodine, which is an indicator of proteins, thus eliminating protein structures. It was determined that the substance belonged to cell nuclei and was actually DNA precipitate from leukocytes [1, 2].

 IntechOpen

After the isolation of DNA or cell nuclei, the properties of the molecule were investigated to determine its role in the cell. The first signature related to heritable material of DNA and its role in determining cell properties came from Griffith's study in 1928. Griffith described a transformation experiment in which pathogenic pneumococcal lysate could influence apathogenic strains, transforming the apathogenic strain into pathogenic strains. However, his study did not actually identify DNA as a molecule that determines cellular properties or transformation.

Avery and colleagues focused on this topic in the 1940s. They showed that when lysate was treated with enzymes that attacked DNA, transformation was inhibited, demonstrating that a heritable substance in cells was DNA [3]. After isolation of DNA and understanding its function in cells, further efforts were related to its composition. Erwin Chargaff filled the gap in science-related DNA composition. Owing to studies by Chargaff, the bases adenine, guanine, cytosine, and thymine, which are the fundamental components of DNA, were well-characterized [4]. After that time, the new questions arose regarding base ordering and genetic code, which opened the door to a new hot topic: DNA sequencing.

The first nucleic acid sequencing was related RNA. RNA sequencing was reported by Robert Holley and his colleagues in 1965. Holley studied alanine transfer RNA and focused on determining the nucleotide base order. He treated alanine tRNA with specific RNA cleavage enzymes including pancreatic ribonuclease and takadiastase ribonuclease. After this specific cleavage of alanine tRNA, the fragments were analyzed by paper chromatography. This study design led him characterize the 77-nucleotide-length alanine transfer RNA sequence [5, 6]. In 1968, Sanger and coworkers used ^{32}P-labeled RNA from ribonuclease digests and the paper fractionation technique to highlight 120-nucleotide-length nucleotide sequence of the 5S ribosomal RNA sequence [7]. In 1969, Sanger and coworkers reported a 57 nucleotide sequence of coat-protein cistron of phage R17 RNA [8]. Following these studies, the complete sequence of two 5s RNA, 16 tRNA, and the long nucleotide order of phage RNA were described [9–11]. After the year 1965, RNA sequencing reports gradually increased. However, no study reported nucleotide order in DNA during those years. The first report to determine nucleotide order in DNA came from Ray Wu's laboratory in 1968. What were the limitations of DNA sequencing? Why did DNA sequencing come later? Several reasons can be listed. First, obtaining homogeneous DNA in a high amount was very difficult. Second, the length of DNA to be obtained is too large compared to RNA, for example, ~50 kb for lambda DNA. Third, highly specific DNases were not available to aid sequence analysis. Fourth, fraction analysis of digested nucleic acids was limited to several base pairs [12–15].

2. DNA sequencing methods in the early years

Elimination of the aforementioned limitations or the development of different approaches enabled the characterization of nucleotide base order in DNA. DNA sequencing techniques evolved significantly in the 1970s, with several scientists focusing on this topic. Since the first DNA sequencing report by Ray Wu in 1968, very significant progress has been recorded in one decade. It is not surprising that previous RNA sequencing methods drove scientists to develop new strategies for sequencing DNA. Similar methods to RNA sequencing or more original methods were developed for sequencing DNA. Two main strategies were adopted for the developed methods. The first was the *sequencing by synthesis* method, in which the nucleotide base order in DNA is determined by synthesizing DNA using DNA polymerase. The second was the *direct sequencing* method, where nucleotide order in the *in vivo* DNA was analyzed directly without synthesis by DNA polymerase. When we look at the DNA sequencing history, it is clearly seen that the year 1977 was a golden year for DNA sequencing. Several scientists were awarded the Nobel Prize. Since reporting of the first DNA sequencing method in 1968, more practical DNA sequencing methods have been described in the first decade, which are highlighted in the following subsections.

2.1 Ray Wu's nucleotide incorporation method

Studies related to the specific digestion of RNA and determination of nucleotide order seem to provide an idea to Wu and coworkers to highlight base order in DNA. It seems that the first DNA sequencing study was described by Ray Wu, a Chinese scientist who contributed to very important studies related to nucleic acid sequence analysis at Cornell University, along with his coworkers [16]. Ray Wu's method involved to *in vitro* synthesis of DNA using labeled nucleotides at the sticky ends of lambda DNA ends and then determining the nucleotide order of the lambda DNA terminus. Lambda DNA ends, which are known as cos sites, enable it to come together to form circular or linear forms (**Figure 1**) [17, 18].

Wu was interested in determining the nucleotide sequence of these cohesive ends [16]. Studying bacteriophage lambda DNA to discover the nucleotide order in lambda DNA termini had two advantages. The first was obtaining homogeneous form in large amounts, and the latter had protruding single-stranded termini (3′) to serve as a primer for DNA polymerase to incorporate labeled deoxynucleotides at low temperatures [14, 16, 19, 20]. In this way, they started experiments. Firstly, they labeled deoxynucleoside triphosphates and *Escherichia coli (E.coli)* DNA polymerase catalyzed the incorporation of nucleotides into lambda DNA *in vitro*, lengthening the 3′-terminated strands. After that, hydrolysis of labeled DNA to nucleoside-3′-monophosphates was carried out by micrococcal DNase, spleen phosphodiesterase, and venom phosphodiesterase and analyzed by paper chromatography (**Figure 2**). This study allowed him to determine the nucleotide composition in the cohesive ends as 20 nucleotides as 13 residues of dG, 13 of dC, 7 of dA, and 7 of dT, indicating that the single-strands protruding ends have complementary base sequences.

In 1970, Wu improved a previous DNA sequence determination method. In the improved method, deoxynucleotides were labeled with radioactive ^{32}P or ^{3}H, and limited or complete incorporation was carried out at the protruding end of

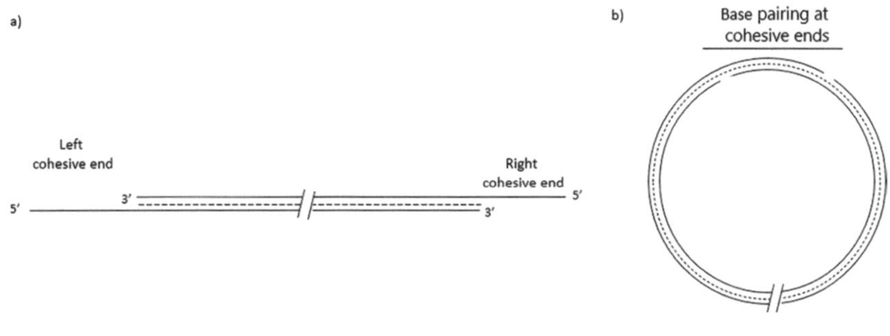

Figure 1.
The forms of bacteriophage lambda (λ) DNA. Right and left cohesive ends are free in the linear form (a). When cohesive ends pairs with hydrogen bonds, the circular form is formed (b).

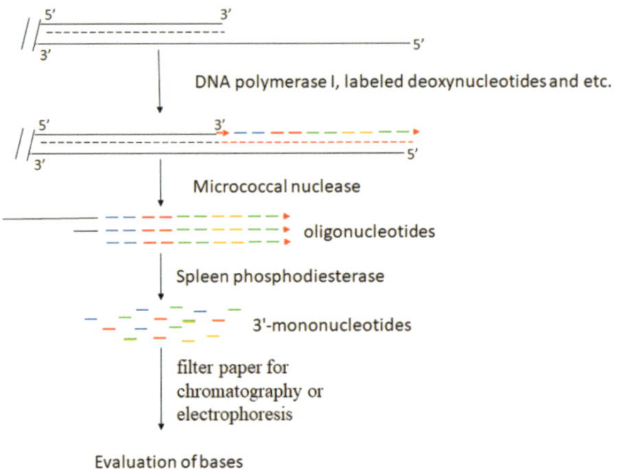

Figure 2.
Wu's nucleotide incorporation method to deduce nucleotide base order in the protruding ends of lambda DNA.

bacteriophage lambda. In the complete incorporation, labeled or repaired DNA was digested partially by micrococcal nuclease. 3′-OH-ended fragments were purified, and pentanucleotides were obtained. Snake venom phosphodiesterase digested the penta-oligonucleotides into shorter ones. Digested nucleotides were electrophoresed on paper and mobility shift was determined. The mobility shift difference showed the nucleotide type in the pentanucleotide, thereby allowing the determination of the 3′-OH nucleotide sequence (**Figure 3**). On the other hand, limited incorporation was carried out to evaluate the 5′ sequence of lambda termini. When only ^{32}P-GTP was added, the polymerization reaction stopped after three guanine additions. He concluded that the first sequence could be GGG. Further polymerization was carried out by adding labeled ^{32}P-GTP and ^{3}H-CTP and radioactivity in the lambda DNA was determined. This limited deoxynucleotide addition and evaluation of radioactivity were performed over a

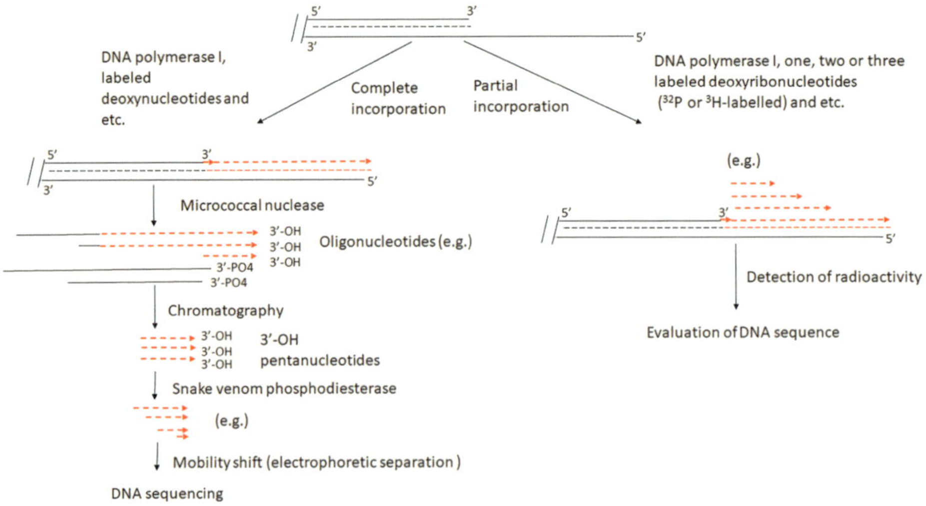

Figure 3.
Wu's improved nucleotide incorporation method to sequence DNA at the end of bacteriophages.

range of combinations, and the evaluation of the results enabled him to determine the 5′ nucleotide sequence of DNA (**Figure 3**). Owing to Wu's strategy, the eight sequences were determined as GGGCGGCG, and the terminal sequence was determined as GCGGCG-OH [14].

Wu showed similar efforts in the new study as in the previous one. The DNA sequence was determined as a 12-nucleotide-length (dodecanucleotid) of GGGCGGCGACCT, starting from the left-hand side, and the sequence was exactly complementary to the right-hand cohesive end. In this report, it was noted that "...*The procedures described here for sequence determination utilizing DNA polymerase-catalyzed incorporation may also be applicable to sequence analysis in other types of DNA molecules*" [21]. This is a very important future aspect that could direct new strategies to sequence another DNA source. In addition to these studies, Wu and his colleagues also highlighted bacteriophage ϕ80 3′ cohesive end sequence by using the described method, in which DNA polymerase I directed the synthesis or labeling of cohesive ends, followed by micrococcal nuclease digestion of labeled DNA and separation of the nucleotides by two-dimensional electrophoresis and sequencing of nucleotides. They found that the cohesive ends of lambda DNA and ϕ80 DNA were homologs to each other [22].

2.2 Englund's T4 DNA polymerase-directed hydrolytic method

Englund [23] designed a study to determine the order of DNA sequence. He described nucleotide sequence analysis using bacteriophage T4-induced DNA polymerase (T4 DNA polymerase) at the 3′ ends of double-strand DNA. The T4 DNA polymerase has hydrolytic activity as well as polymerization activity. The main mechanism of Englund' s approach was that a single triphosphate causes the release of one nucleotide from each strand, determining nucleotide sequence at the 3′ ends of ^{32}P-labeled duplex DNA. For example, in the presence of dCTP, T4 DNA polymerase begins to degrade at the 3′ ends of double-strand DNA until it reaches the cytosine base in the strand. Other bases are released from double-strand DNA (**Figure 4**). The released nucleotide monophosphates were determined by thin-layer chromatography, and the order of the nucleotide bases was determined. The method was tested on

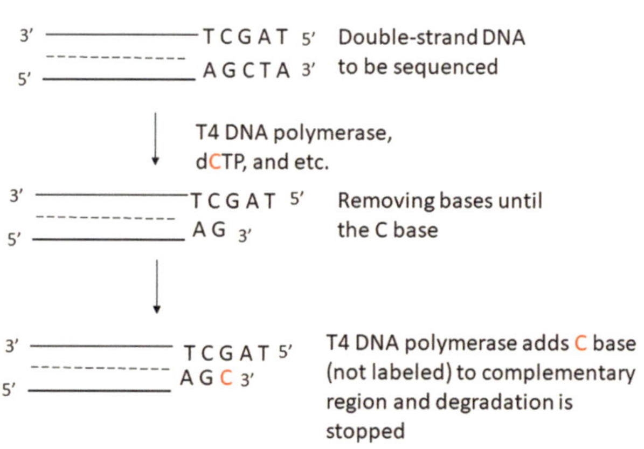

Figure 4.
DNA sequencing analysis at 3′ termini of dsDNA using the T4 DNA polymerase-directed hydrolytic method by Paul T. Englund.

Hemophilus influenzae Endoculease R-treated DNA, and it was able to determine the nucleotide sequence at the end of DNA [23].

2.3 Wu's primer-extension method

Until 1972, Ray Wu and his coworker had used self-priming of lambda DNA protruding ends and attempted to discover nucleotide sequence order in the phage. In 1972, Ray Wu developed the method by using oligonucleotides as a primer to sequence a specific DNA region [24, 25]. Padmanabhan et al. [24] synthesized a radioactive oligonucleotide primer, which contained pyrimidine octamers with the sequence d(C-T-T-T-C-C-C-C), by using *E. coli* DNA polymerase and catalyzing repair on the phage 186 DNA. After the reaction, the synthesized oligonucleotides were purified by electrophoresis at pH 3.5 on DEAE-cellulose paper to be used as primers in their study (**Figure 5**). Using the purified radioactive octanucleotide, which was eight nucleotides in length, they showed the octanucleotide served as a primer to synthesize by DNA polymerase, although the octanucleotide did not form a stable duplex with 186 DNA (**Figure 6**). They also reported that adding one, three, or eight nucleotides of the defined primer increased the likelihood of duplex formation, in other words, binding to the target DNA sequence. In the report, they note that "*it will be a powerful tool for future DNA sequence analysis*," which is a very important future aspect because the study opened the door to primer-directed DNA sequencing studies.

Wu and colleagues improved the primer-extension method by using a chemically synthesized oligonucleotide with a length of 12 nucleotides (dodecadeoxynucleotide) as a primer to bind to the endolysin gene of lambda DNA for sequencing the gene. Until that time, they had used self-priming and sequencing bacteriophage terminal sequences; however, in the new approach, they aimed to sequence specific regions,

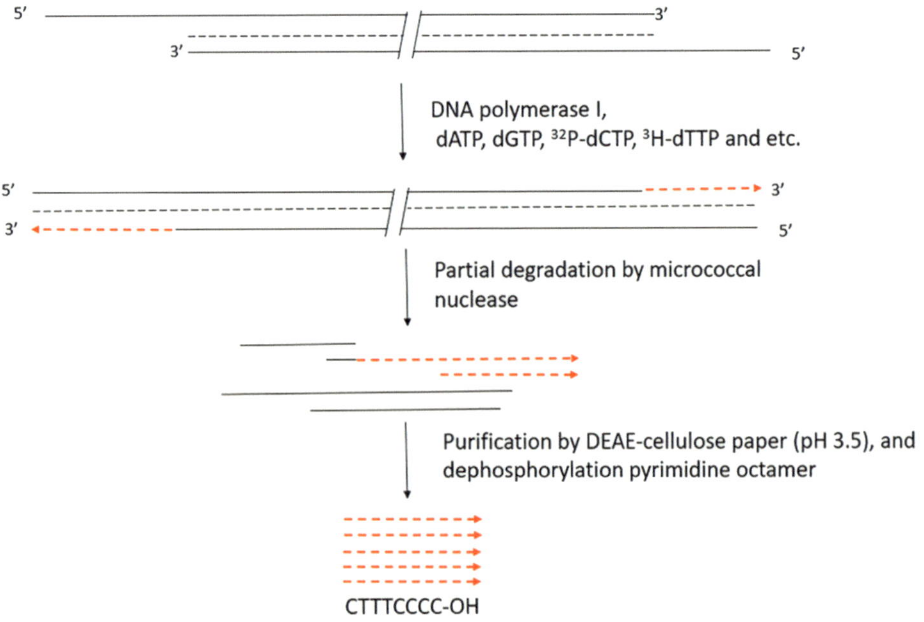

Figure 5.
The synthesis and purification of the pyrimidine octamer primer by the end repair method using DNA polymerase I.

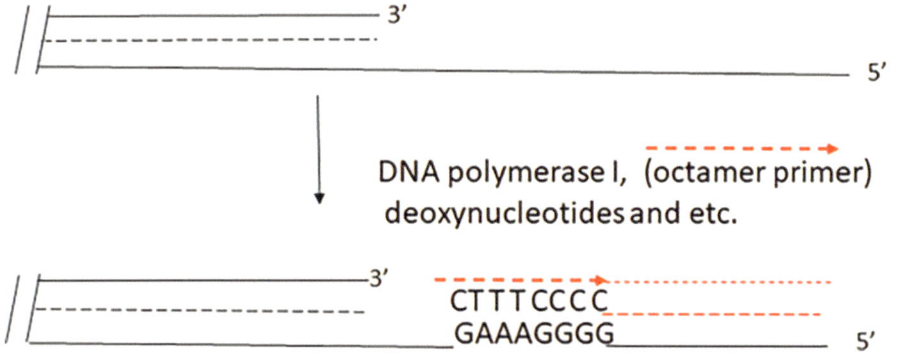

Figure 6.
Synthesis and sequencing lambda termini by DNA polymerase I directed primer (pyrimidine octamer).

which was the endolysin gene. The endolysin gene was chosen because its location is close to the lambda termini and the sequence of the endolysin protein is known. Four amino acid sequences (132-Gln-Phe-Glu-His-137) were selected in the endolysin protein sequence to design the primer sequence. Based on the known protein sequence, the mRNA sequence was estimated, which determined the DNA sequence as 5′-CAG TTT GAG CAT-3′ and the sequence was selected as a primer and synthesized chemically (**Figure 7**). Their optimized method highlighted that the dodeca-deoxynucleotide initiated nucleotide incorporation at its 3′ ends, and the technique enabled the sequencing of a specific region of DNA (**Figure 8**) [26]. This approach was also very important, as it opened new doors in DNA sequencing, and oligonucleotides could serve as primers for synthesizing DNA and sequencing approaches. Even in current studies, oligonucleotides of approximately 20 nucleotides in length are usually used as primers to start DNA synthesis *in vitro*.

2.4 Salser's ribosubstitution method

An important limitation of DNA sequencing was the lack of a specific cleavage enzymes compared to RNA sequencing. For example, T1 RNase was used in RNA sequencing to hydrolyze RNA specifically at Gp residues. Salser et al. [27] proposed

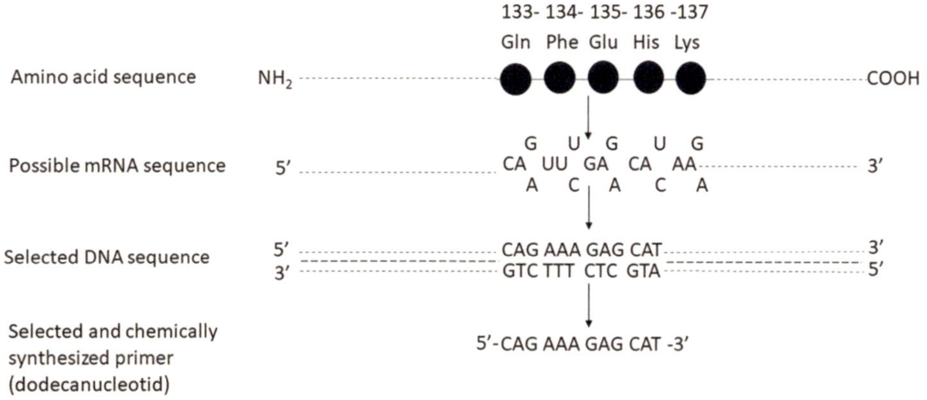

Figure 7.
Primer selection strategy for sequencing the defined endolysin gene region of the bacteriophage lambda.

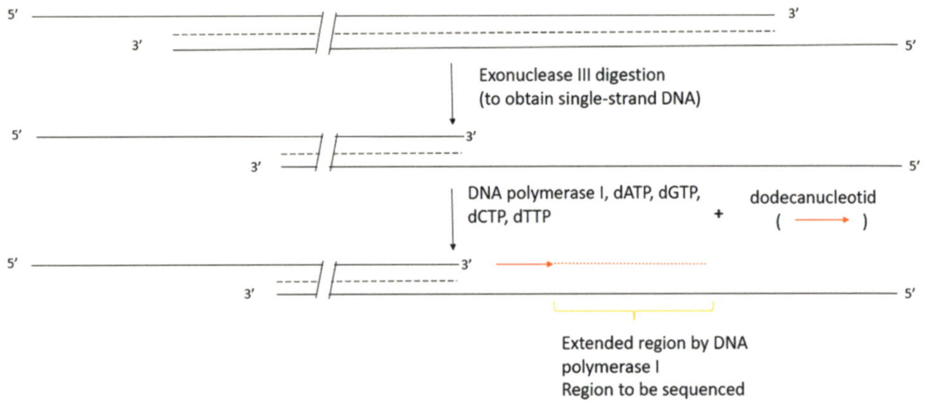

Figure 8.
The improved primer-extension method for DNA sequencing developed by Wu. Specific region sequencing using chemically synthesized primers and DNA polymerase to extend the DNA region to be sequenced.

that ribonucleotides could be incorporated into DNA, allowing enzymatic or chemical breakage at these ribonucleotides, thus enabling DNA sequencing. This idea was based on the fact that when *in vitro* DNA synthesis is carried out in the presence of Mn^{++} instead of magnesium ions, ribonucleotides can be incorporated into the newly synthesized DNA. Since RNA-specific cleavage agents are well-known, incorporating ribonucleotides into DNA and achieving base-specific cleavage by RNase or chemicals is an important attempt to overcome this limitation. Salser et al. [27] synthesized DNA *in vitro* in the presence of Mn^{2+}, successfully incorporating rGTP. Alkaline hydrolysis cleaved specifically at the ribose linkage in every G region, allowing sequencing of DNA. The ribosubstitution method allowed site-specific cleavage either at guanine, adenine, or cytosine residues, which let them to the analysis of the nucleotide sequence.

2.5 Sanger's ribonucleotide or partial incorporation method

Sanger et al. [28] focused on the DNA sequence analysis of a specific DNA region by using synthetic primers. Their method was similar to the method described by Salser et al. [27]. Sanger et al. [28] used chemically synthesized octadeoxyribonucleotide (8 nucleotide length) to initiate DNA synthesis by *E. coli* DNA polymerase I in the presence of several conditions. The main aim was that *E. coli* DNA polymerase incorporates ribonucleotide (rCTP) to DNA in the presence of manganese ions, and treatment of DNA with ribonuclease A cleaves only the rC residue. Another condition was that one or more deoxynucleotide triphosphate was added to the reaction in the presence of magnesium ions, which showed similarity to Wu's partial incorporation method. Using the two-dimensional homochromatography technique, they investigated partial degradation products to determine nucleotide sequence. In the study, Sanger obtained the first evidence that chain termination in the absence of a deoxynucleotide triphosphate could be used for sequencing strategy. In the absence of a deoxynucleotide triphosphate, DNA polymerase could not synthesize the DNA to the complementary sequence; this finding could be an important development of minus and chain termination-based DNA sequencing methods.

2.6 Ziff et al.' direct DNA sequencing approach

The main progress in DNA sequencing came from *in vitro* synthesized DNA studies mentioned above. However, the direct sequencing approach of *in vivo* DNA was a special topic since *in vitro* synthesized DNA could bear enzyme-directed incorrect base determination and more complex processes were needed. In 1973, Ziff et al. [15] studied to find ways to improve the direct sequencing of DNA from *in vivo* sources. Since *in vivo* DNA structures were generally larger, purify, and sequencing difficulties were arising. Therefore, they used T4 phage-induced endonuclease IV to obtain partial digests. *In vivo* synthesized DNA was treated by T4 phage-induced endonuclease IV to obtain partial digests and DNA fragments were obtained for further degradation. The obtained DNA fragments were treated by depurination and partial digestion by exonuclease treatment such as spleen and snake venom phosphodiesterase. It was shown that the method was particularly applicable to DNA in combination with several methods and could be used to sequence DNA fragments from endonuclease IV digestion [15, 29].

2.7 Sanger's plus and minus method

In 1975, Sanger described a new method called as "plus and minus" method [30]. Findings from studies by R. Wu and P.T. Englund, played an important role in the development of Sanger's "minus and plus" method to deduce nucleotide sequence in DNA [16, 21, 23, 30, 31]. In this method, *in vitro* DNA synthesis was carried out by using a primer, DNA polymerase, and four types of deoxynucleotide triphosphates (dNTPs). The reaction was ended at different time intervals, and primes with different lengths were obtained. The extended primers were purified and then used in the following two reactions. In the minus reaction, the purified and extended primer, which was the random mixture of oligonucleotides, was reincubated to initiate synthesis by DNA polymerase I in the presence of three deoxyribotriphosphates. If dATP was missing (– A system), DNA polymerase would terminate the reaction before A residue. This system was carried out separately for T, G, and C bases. Thus, four minus reactions were carried out. In the other step, the random mixture of oligonucleotides was treated by T4 DNA polymerase, which had been obtained from T4-infected *E. coli*, each tube containing a triphosphate to inhibit DNA cleavage on 3′ ends. For example, if a DNA was ended A, T4 DNA polymerase did not cleavage the DNA fragment in deoxyadenine triphosphate-containing tubes (+ A system). Four types plus reactions were carried out. The obtained DNA fragments from eight tubes were resolved by a 12% polyacrylamide gel (**Figure 9**). Then, the bands obtained from the tubes were compared and read to deduce nucleotide sequence order up to 50 bases [30]. This system eliminated the use of specific DNA cleavage enzymes or chemicals. After incubation, DNA fragments directly fragmented in the polyacrylamide gel and nucleotide order could be read directly on the gel.

2.8 Maxam and Gilbert's chemical cleavage method

In 1977, Maxam and Gilbert [32] reported a new DNA sequencing method that was based on the specific chemical cleavage of DNA. In the method, *in vivo* DNA can be sequenced directly; therefore, the method has special importance due to a part of the *direct DNA sequencing* method. The method included labeling of DNA by ^{32}P and then

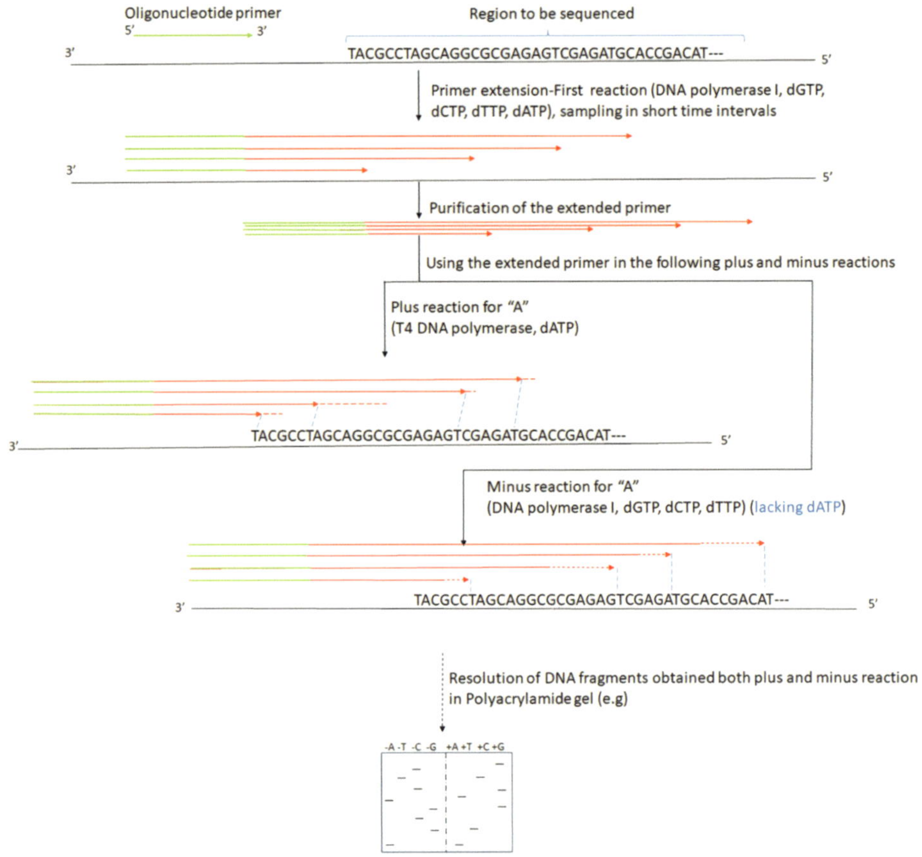

Figure 9.
Sanger's plus and minus method in which only A situation is shown. Similar biological reactions are carried out for the other nucleobases.

three steps including base modification in DNA, cleavage of the modified base, and breaking DNA from the remaining sugar position were carried out. After cleavage of DNA, fragments were separated by polyacrylamide gel and determined autoradiography owing to ^{32}P labeling. In the method, the most important strategy was the specific cleavage of DNA by chemical modification or attraction. To break DNA specifically on guanine and adenine bases, DNA was treated by dimethyl sulfate and heat in natural pH, whereas dimethyl sulfate and dilute acid treatment caused adenine-specific cleavage. Hydrazine and piperidine treatment break DNA on cytosine and thymine bases. Also, hydrazine treatment was carried out in the presence of salt to break cytosine bases specifically. For this purpose, the four conditions were applied separately in the four tubes. When the application was finalized, tubes were loaded with polyacrylamide gel, and the obtained fragments were separated. Lastly reading of the gel was performed from bottom to top, thereby DNA sequence was determined (**Figure 10**). This strategy enabled the sequence of a 64-basepair DNA fragment of lac operon in 1977 [32].

2.9 Sanger's chain termination method

Sanger's previously described DNA sequencing methods had some limitations such as labor. Furthermore, more advantageous DNA sequencing was described

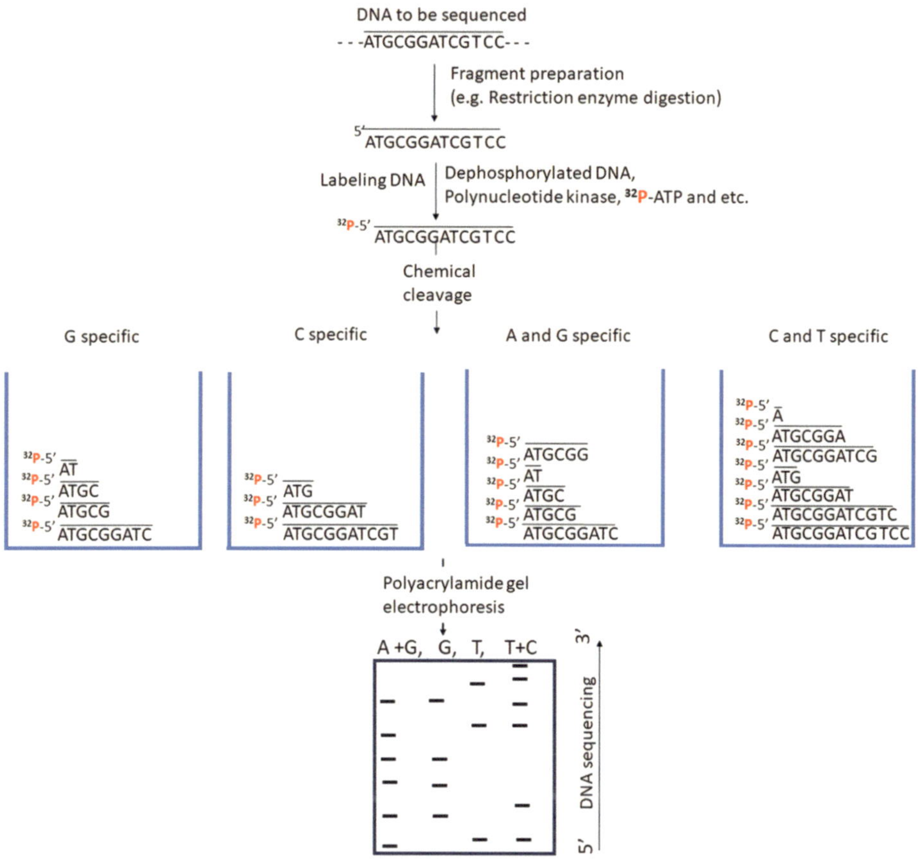

Figure 10.
The DNA sequencing method based on specific chemical cleavage developed by Maxam-Gilbert in 1977.

by Maxam-Gilbert. Therefore, Sanger focused on the more easy or advantageous method. In previous studies, he used chain termination by adding no specific nucleotides. He realized that 2′,3′-dideoxysthymidine (ddTTP) can incorporated into the newly synthesized DNA instead of dT, and ddTTP lacks free 3′ end, this leads to chain termination at specific dT sites, inhibiting DNA polymerase I depending on the specific region. Sanger thought that this chain termination effect could be used for all other deoxyribonucleoside triphosphates. When primer extension was carried out by DNA polymerase I in the presence of all deoxyribonucleoside triphosphates and also ddTTP, different lengths of fragments were obtained and all have the same 3′ ends where ddTTP. Therefore, in the template DNA, dT location could be obtained. This strategy was performed for all four bases with dideoxy nucleotide in a separate way. The fractions were resolved by denaturating polyacrylamide gels and DNA sequencing was identified (**Figure 11**). In the method, starting from the primer site 15 or 200 base sequences could be determined with desired accuracy. It was also possible to up to 300 bases. More length DNA sequence could be determined by the method, however; gel resolution was stringing it.

It should be noted that Sanger's dideoxy sequencing method required single-strand DNA to bind and extend of primer. Several approaches were developed to overcome this limitation and those approaches improved the applicability of the

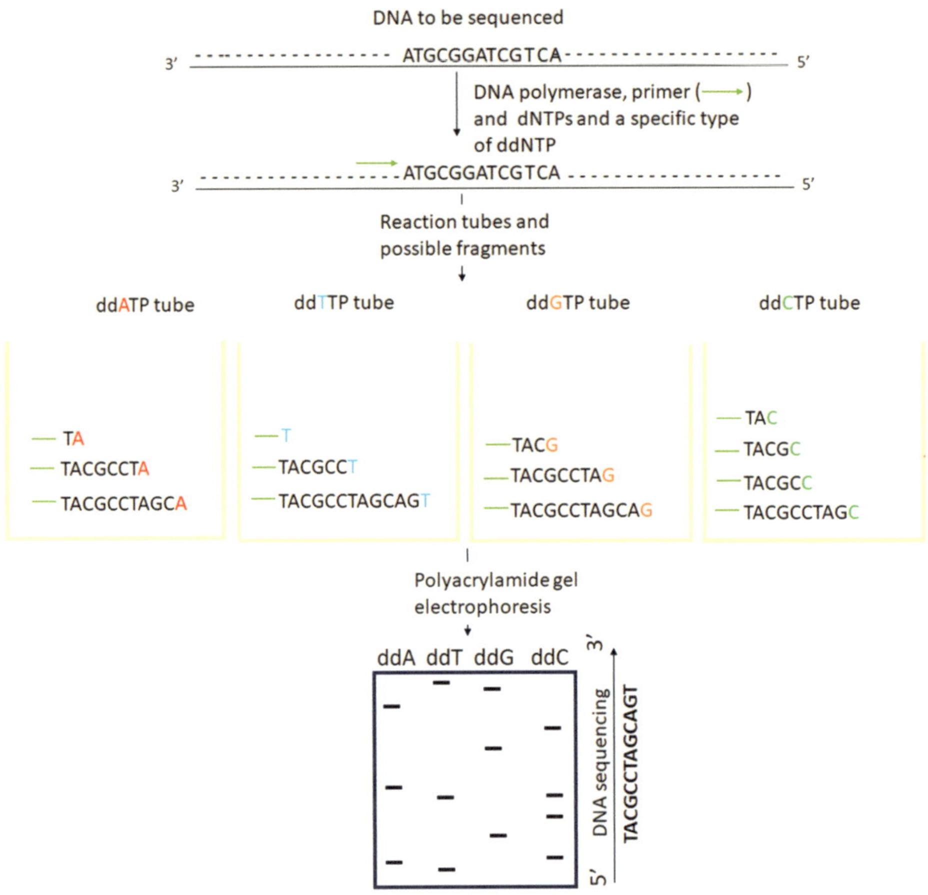

Figure 11.
Sanger's chain-termination DNA sequencing method, introduced in 1977, is based on the incorporation of dideoxynucleoside into the newly synthesized strand.

dideoxy sequencing method. In 1978, Maat and Smith [33] described limited nicking of double-strand DNA by using pancreatic DNAase I and following extension of 3′-OH nicked ends by DNA polymerase I. In another approach, double-strand DNA to be sequenced was cloned into bacteriophage vectors with single-strand such as ϕXI74 or developed vectors, for example, M13 [34, 35]. Using the flanking primer cloned DNA fragment was sequenced on the single-stranded vector. On the other hand, Smith [36] used the Exonuclease III to obtain a suitable template for the dideoxy sequencing method. The method was developed based on the fact that Exonuclease III shows exonuclease activity at the end of 3′-OH in the double-strand DNA, giving partially single-strand DNA. Therefore, the improved method enabled to sequencing be more practical, thereby being preferred by a lot of scientists.

3. Conclusion

Starting from 1968, the first DNA sequencing affords, DNA sequencing methods improved fascinating fast. Only one decade was enough to reach to top methods.

Even though the first methods showed similarities to RNA sequencing methods, more creative ones were developed to overcome the difficulty of DNA sequencing. In the sequencing strategies mainly two approaches were developed; *sequencing by synthesis* and *direct sequencing*. The first scientist to introduce *sequencing by synthesis* was Wu in 1968. On the other hand, *direct sequencing* was described by Englund in 1971. Therefore, DNA sequencing strategies were developed in two ways after 1971 (**Figure 12**). Developments of the methods in the timeline are summarized in **Figure 12**. Maxam-Gilbert's sequencing method was the top of *direct sequencing* in 1977. On the other hand, Sanger's dideoxy chain termination method was the more practical method of *sequencing by synthesis* (**Figure 12**). Even though these DNA sequencing methods were very useful and empowering genetics studies, Ray Wu's contributions were not ignorable. Even though the Nobel Prize was not given to Ray Wu, DNA sequencing history was started with studies of him and his coworkers. Partial nucleotide incorporation and primer-extension methods for DNA sequencing were first described by Wu.

It seems that polyacrylamide gel electrophoresis, also called slab gel electrophoresis, empowered Sanger and Maxam-Gilbert's studies. However, Wu did not use the slab gel electrophoresis system. He focused on the digestion of DNA and two-dimensional paper chromatography techniques which were difficult for others.

It should be noted that one strategy for DNA sequencing includes *in vitro* transcription. DNA to be sequenced was converted RNA by *in vitro* transcription to obtain complementary RNA (cRNA), and RNA sequencing was achieved owing to high specific cleavage enzymes such as ribonuclease T1 (for example [37, 38]). However, in this chapter direct DNA-related methods were highlighted.

To sum up, the elimination of nucleases, development of gel electrophoresis, and/or obtaining specific DNA cleavage techniques increased the applicability of DNA sequencing. These improvements were at the core of DNA sequencing in 1977. Since the more advantageous *direct sequencing* and *sequencing by synthesis* techniques were

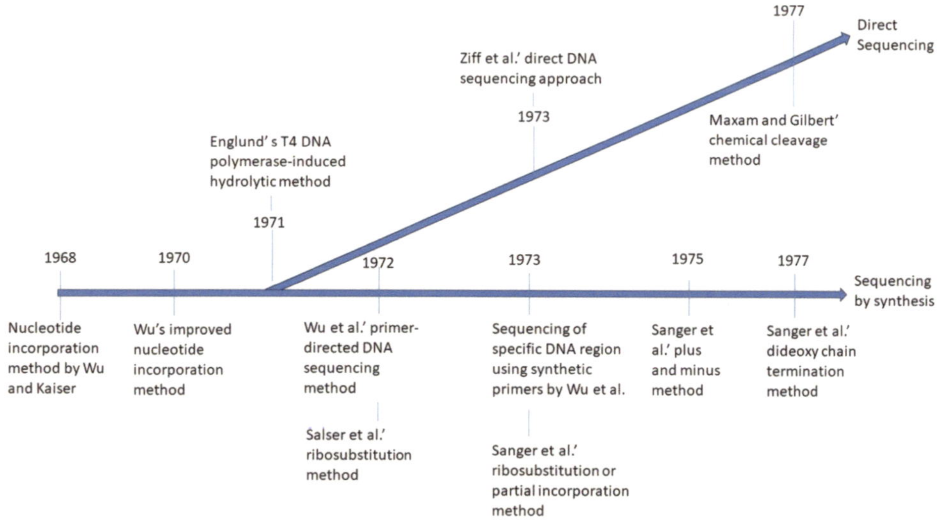

Figure 12.
Timeline of the developments in DNA sequencing methods during the first decade.

introduced in 1977, the year should be the golden age in DNA sequencing history. Even though Sanger's dideoxy method is about 50 years old, the method is still the preferred method today.

Conflict of interest

The author declares no conflict of interest.

Author details

Mevlüt Arslan
Department of Genetics, Faculty of Veterinary Medicine, Van Yüzüncü Yıl University, Tuşba, Van, Turkey

*Address all correspondence to: mevlutarslan@yyu.edu.tr

References

[1] Miescher JF. Ueber die Chemische Zusammensetzung der Eiterzellen. Medicinisch-Chemische Untersuchungen. 1871;**4**:441-460

[2] Wolf G. Friedrich Miescher, the man who discovered DNA. Chemical Heritage. 2003;**21**(10-11):37-41

[3] Avery OT, Macleod CM, McCarty M. Studies on the chemical nature of the substance inducing transformation of pneumococcal types: Induction of transformation by a desoxyribonucleic acid fraction isolated from pneumococcus type iii. The Journal of Experimental Medicine. 1944;**79**(2):137-158

[4] Chargaff E. Chemical specificity of nucleic acids and mechanism of their enzymatic degradation. Experientia. 1950;**6**(6):201-209

[5] Holley RW. Structure of an alanine transfer ribonucleic acid. Journal of the American Medical Association. 1965;**194**(8):868-871

[6] Holley RW, Apgar J, Everett GA, Madison JT, Marquisee M, Merrill SH, et al. Structure of a ribonucleic acid. Science. 1965;**147**(3664):1462-1465

[7] Brownlee G, Sanger F, Barrell B. The sequence of 5S rRNA. Journal of Molecular Biology. 1968;**34**:379-412

[8] Adams JM, Jeppesen PGN, Sanger F, Barrell BG. Nucleotide sequence from the coat protein cistron of R17 bacteriophage RNA. Nature. 1969;**223**(5210):1009-1014

[9] Madison JT. Primary structure of RNA. Annual Review of Biochemistry. 1968;**37**:131-148

[10] Zachau HG. Transfer ribonucleic acids. Angewandte Chemie, International Edition. 1969;**8**(10):711-727

[11] Billeter MA, Dahlberg JE, Goodman HM, Hindley J, Weissmann C. Sequence of the first 175 nucleotides from the 5′ terminus of Qbeta RNA synthesized in vitro. Nature. 1969;**224**(5224):1083-1086

[12] Robertson HD, Barrell BG, Weith HL, Donelson JE. Isolation and sequence analysis of a ribosome-protected fragment from bacteriophage φX 174 DNA. Nature: New Biology. 1973;**241**(106):38-40

[13] Ling V. Fractionation and sequences of the large pyrimidine oligonucleotides from bacteriophage fd DNA. Journal of Molecular Biology. 1972;**64**(1):87-102

[14] Wu R. Nucleotide sequence analysis of DNA. I. Partial sequence of the cohesive ends of bacteriophage lambda and 186 DNA. Journal of Molecular Biology. 1970;**51**(3):501-521

[15] Ziff EB, Sedat JW, Galibert F. Determination of the nucleotide sequence of a fragment of bacteriophage phiX 174 DNA. Nature: New Biology. 1973;**241**(106):34-37

[16] Wu R, Kaiser AD. Structure and base sequence in the cohesive ends of bacteriophage lambda DNA. Journal of Molecular Biology. 1968;**35**(3):523-537

[17] Hershey AD, Burgi E, Ingraham L. Cohesion of DNA molecules isolated from phage lambda. Proceedings of the National Academy of Sciences of the United States of America. 1963;**49**(5):748-755

[18] Machattie LA, Thomas CA Jr. DNA from bacteriophage lambda: Molecular length and conformation. Science. 1964;**144**(3622):1142-1144

[19] Richardson CC, Inman RB, Kornberg A. Enzymic synthesis of deoxyribonucleic acid. 18. The repair of partially single-stranded DNA templates by DNA polymerase. Journal of Molecular Biology. 1964;**9**:46-69

[20] Wu R, Kaiser AD. Mapping the 5′-terminal nucleotides of the DNA of bacteriophage lambda and related phages. Proceedings of the National Academy of Sciences of the United States of America. 1967;**57**(1):170-177

[21] Wu R, Taylor E. Nucleotide sequence analysis of DNA. II. Complete nucleotide sequence of the cohesive ends of bacteriophage lambda DNA. Proceedings of the National Academy of Sciences of the United States of America. 1967;**57**(1):170-177

[22] Bambara R, Padmanabhan R, Wu R. Nucleotide sequence analysis of DNA. X. Complete nucleotide sequence of the cohesive ends of bacteriophage phi80 DNA. Journal of Molecular Biology. 1973;**75**(4):741-744

[23] Englund PT. Analysis of nucleotide sequences at 3′ termini of duplex deoxyribonucleic acid with the use of the T4 deoxyribonucleic acid polymerase. The Journal of Biological Chemistry. 1971;**246**(10):3269-3276

[24] Padmanabhan R, Padmanabhan R, Wu R. Nucleotide sequence analysis of DNA. IX. Use of oligonucleotides of defined sequence as primers in DNA sequence analysis. Biochemical and Biophysical Research Communications. 1972;**48**(5):1295-1302

[25] Wu R. Nucleotide sequence analysis of DNA. Nature: New Biology. 1972;**236**(68):198-200

[26] Wu R, Tu CD, Padmanabhan R. Nucleotide sequence analysis of DNA. XII. The chemical synthesis and sequence analysis of a dodecadeoxynucleotide which binds to the endolysin gene of bacteriophage lambda. Biochemical and Biophysical Research Communications. 1973;**55**(4):1092-1099

[27] Salser W, Fry K, Brunk C, Poon R. Nucleotide sequencing of DNA: Preliminary characterization of the products of specific cleavages at guanine, cytosine, or adenine residues (bacteriophage M13-ribosubstitution-DNA polymerase I-electrophoresis-two-dimensional fingerprinting). Proceedings of the National Academy of Sciences of the United States of America. 1972;**69**(1):238-242

[28] Sanger F, Donelson JE, Coulson AR, Kössel H, Fischer D. Use of DNA polymerase I primed by a synthetic oligonucleotide to determine a nucleotide sequence in phage f1 DNA. Proceedings of the National Academy of Sciences of the United States of America. 1973;**70**(4):1209-1213

[29] Galibert F, Sedat J, Ziff E. Direct determination of DNA nucleotide sequences: Structure of a fragment of bacteriophage phiX172 DNA. Journal of Molecular Biology. 1974;**87**(3):377-407

[30] Sanger F, Coulson AR. A rapid method for determining sequences in DNA by primed synthesis with DNA polymerase. Journal of Molecular Biology. 1975;**94**(3):441-448

[31] Englund PT. The 3′-terminal nucleotide sequences of T7 DNA. Journal of Molecular Biology. 1972;**66**(2):209-224

[32] Maxam AM, Gilbert W. A new method for sequencing DNA. Proceedings of the National Academy of

Sciences of the United States of America. 1977;**74**(2):560-564

[33] Maat J, Smith AJH. A method for sequencing restriction fragments with dideoxynucleoside triphosphates. Nucleic Acids Research. 1978;**5**(12):4537-4546

[34] Gronenborn B, Messing J. Methylation of single-stranded DNA in vitro introduces new restriction endonuclease cleavage sites. Nature. 1978;**272**(5651):375-377

[35] Barnes WM. Construction of an M13 histidine-transducing phage: A single-stranded cloning vehicle with one EcoRI site. Gene. 1979;**5**(2):127-139

[36] Smith AJH. The use of exonuclease III for preparing single stranded DNA for use as a template in the chain terminator sequencing method. Nucleic Acids Research. 1979;**6**(3):831-848

[37] Marotta CA, Forget BG, Weissman SM, Verma IM, McCaffrey RP, Baltimore D. Nucleotide sequences of human globin messenger RNA. Proceedings of the National Academy of Sciences of the United States of America. 1974;**71**(6):2300-2304

[38] Gilbert W, Maxam A. The nucleotide sequence of the lac operator. Proceedings of the National Academy of Sciences of the United States of America. 1973;**70**(12):3581-3584

Section 2

DNA Sequencing Applications and Use

Chapter 3

DNA Sequencing Technologies in Accelerating Molecular Breeding

Rhitisha Sood and Vivek Singh

Abstract

A fundamental understanding of DNA structure and function has contributed significantly to our knowledge of genetics. The DNA sequence is a key factor that underlies all inherited traits, making DNA sequence analysis a powerful tool for studying genetics. DNA sequencing allows researchers to determine the base sequence of DNA found in genes and other chromosomal regions. It is one of the most critical methods for exploring genetics at the molecular level. It has become indispensable for basic biological research and various applied fields, including biotechnology, forensic biology, and biological systematics. Molecular geneticists frequently employ DNA sequencing to determine DNA base sequences as a first step toward understanding gene expression and function. For instance, investigating genetic sequences has helped elucidate promoters' function, regulatory elements, and the genetic code. Similarly, analyzing sequences has facilitated our understanding of the origins of replication, centromeres, telomeres, and transposable elements. This chapter will provide beneficial details to researchers and readers with access to advancements and various applications related to sequencing technologies.

Keywords: DNA, fingerprinting, genome, molecular, sequencing

1. Introduction

Genome sequencing in plants is the process of determining the complete DNA sequence of a plant's genome. The genome of a plant contains all the genetic information necessary for its development, growth, and reproduction. Genome sequencing allows researchers to identify and analyze specific genes, study the relationships between different plant species, and understand the genetic basis of plant traits such as disease resistance, stress tolerance, and yield [1]. There are several methods of genome sequencing, but the most common technique involves breaking the genome into small fragments and sequencing them using high-throughput DNA sequencing technologies [2]. The resulting data is then assembled into a contiguous sequence using specialized software. Genome sequencing has become increasingly important in plant research because it provides comprehensive and accurate genetic information for a given species. This information can be used to study plant evolution, identify genetic markers for breeding and trait improvement, and understand the mechanisms underlying plant–environment interactions. Several methods of genome sequencing have been developed over the years, such as first-generation, which has been replaced by next-generation, and third-generation sequencing, which are faster and cost-effective. However, the choice of genome sequencing method depends on factors such as the size and complexity of the genome, the level of accuracy required, and the cost and throughput of the sequencing platform. Each method has its strengths and weaknesses, and researchers often choose to use a combination of methods to generate the most accurate and comprehensive genome sequence possible. The availability of genome sequencing data has also facilitated the development of new technologies such as CRISPR-Cas gene editing, which allows for precise manipulation of specific genes within a plant's genome. Overall, genome sequencing has become a powerful tool for plant scientists and has revolutionized the field of plant biology.

2. History

The Maxam-Gilbert method, which was followed by the Sanger method, was developed in the 1970s and marked the beginning of DNA molecule sequencing. Most DNA sequencing nowadays is done in medical and scientific labs utilizing sequencers using versions of the Sanger technique, first devised by Frederick Sanger in 1975. Researchers could sequence only a modest number of base pairs using the very first DNA sequencing techniques developed in the 1970s. The situation had slightly improved by 1990, but a limited number of laboratories could still sequence (100,000 bases). Additionally, the sequencing process itself was quite expensive and unworkable.

Fortunately, a lot has changed since then, especially in terms of technical development. Even better, automation has made the process much more efficient and useful. Individual genes may now be sequenced swiftly and inexpensively in laboratories regularly. In some labs, more than 100 million bases are sequenced annually (**Figure 1**).

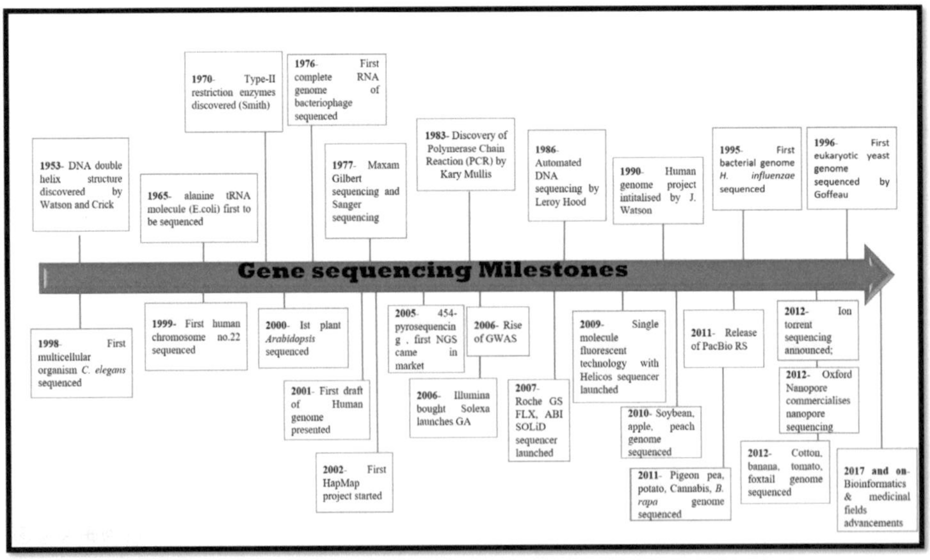

Figure 1.
Timeline of the era of gene sequencing.

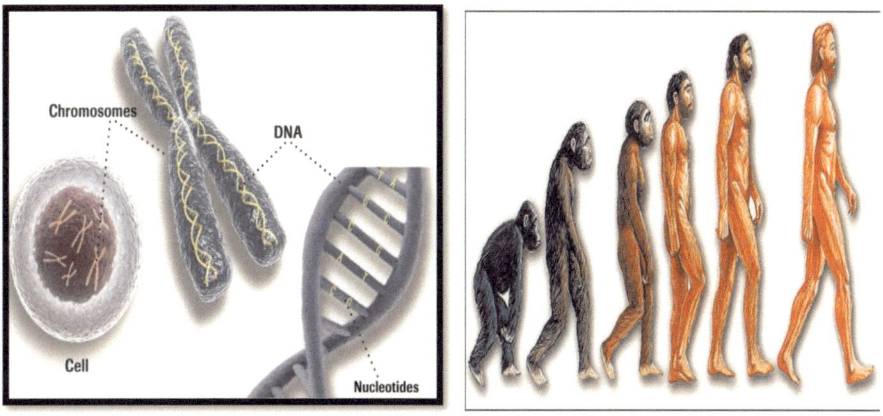

Figure 2.
Depiction of chromosome and DNA, left; evolution of man, right.

2.1 How is DNA sequencing performed?

Finding the exact placement of the four bases found in a single piece of DNA is the process of sequencing DNA (**Figure 2**). Actually, the DNA serves only as a template for the production of many pieces. The fragments are divided by size before the bases are determined, effectively recreating the original DNA sequence. The fragments differ in length by one base. One copy of the human genome, or 23 pairs of chromosomes, is present in each individual. We are restricted in how many bases can be read simultaneously due to technological limitations. As a result, we cannot simply read each base along a chromosome. The chromosome is divided into smaller pieces to make it viable.

3. Types

3.1 1st generation sequencing

3.1.1 Maxam-Gilbert sequencing

In 1977, Allan Maxam and Walter Gilbert published a DNA sequencing technique based on chemically altering DNA and subsequent base cleavage (**Table 1**). This technique, often called chemical sequencing, made it possible to use double-stranded DNA materials purified without further cloning. After the Sanger methods had been improved, their extensive use was discouraged due to the required radioactive tagging and technical difficulty. This procedure begins with a single DNA strand with radioactive ^{32}P attached to one end (5′ or 3′) (**Figures 3** and **4**).

In the four reactions (G, A + G, C, and C + T), the chemical treatment causes breakage at a minor amount of one or two of the four nucleotide bases. For instance, the guanines (and to some extent the adenines) are methylated by dimethyl sulfate, the guanines (A + G) are depurinated using formic acid, and the pyrimidines (C + T) are hydrolyzed using hydrazine. Thymine reaction for the C-only reaction is inhibited by adding salt (sodium chloride) to the hydrazine reaction. Hot piperidine; $(CH_2)5NH$ may then be used to cleave the changed DNAs at the base's location. In order to introduce an average of one change per DNA molecule, the concentration of the modifying agents is

	First generation	Second generation	Third generation
Fundamental technology	Size separation of specifically end labeled DNA fragments	Wash and scan SBS	Single molecule real time sequencing
Resolution	Averaged across many copies of DNA molecule	Averaged across many copies of DNA molecule	Single DNA molecule
Current raw read accuracy	High	High	Lower
Current read length	Moderate (800-1000bp)	Short (generally much shorter than Sanger's sequencing)	>1000 bp
Current throughput	Low	High	High
Current cost	High cost per base	Low cost per base	Low cost per base
	Low cost per run	High cost per run	High cost per run
RNA sequencing method	cDNA sequencing	cDNA sequencing	Direct RNA sequencing
Result timings	Hours	Days	<1 day
Sample preparation	Moderately complex (No PCR amplification required)	Complex (required PCR amplification)	Various
Data analysis	Routine	Complex (due to large data vol. and short reads)	Complex
Primary results	Base calls with quality values	Base calls with quality values	Base calls with quality values
Schdat et al. (2010)			

Table 1.
Table highlighting the differences among three different generations based sequencing.

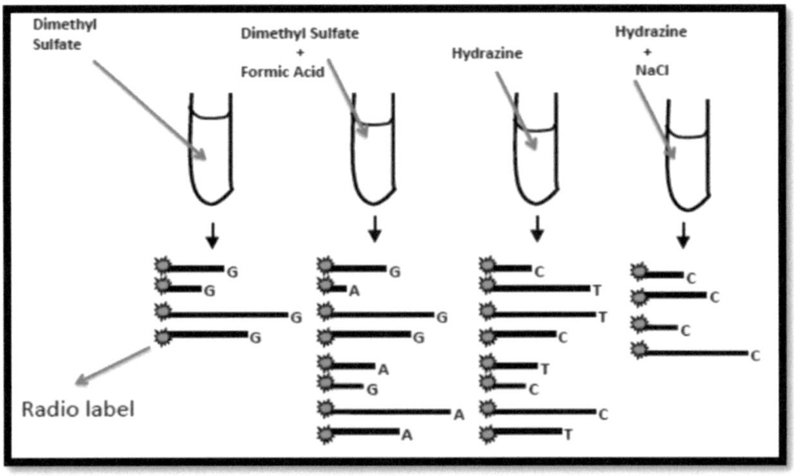

Figure 3.
Radio-labeling of bases.

Figure 4.
Differentiation among gene sequences of human and chimpanzee at three bases.

adjusted. As a result, each molecule produces a succession of labeled fragments starting at the radiolabeled end and ending at the first "cut" site. For size separation, the fragments from the four processes are electrophoresed side by side in denaturing acrylamide gels. When the gel is exposed to X-ray film for autoradiography in order to see the fragments, a sequence of dark bands indicating the locations of identical radiolabeled DNA molecules is produced. The sequence can be deduced from the presence and absence of specific elements.

3.1.2 Sanger's method of gene sequencing

Frederick Sanger and his team created it in 1977, and it won the Nobel Prize in 1980. Dideoxy chain termination method is another name for Sanger's method of gene sequencing. It is a technique for determining the nucleotide sequence of an unidentified DNA strand. For large-scale genome investigations and for getting particularly lengthy DNA sequence reads (>500 nucleotides), Sanger sequencing has more recently been updated as the "Next-Generation sequencing method."

3.1.2.1 Basic principle

Terminators (di-deoxynucleotides), which are frequently utilized in in vitro DNA synthesis processes, cause the growing DNA strand to come to an exact end at a certain location. After the procedure, the strands are overlapped to create the DNA strand's original sequence. It creates a nested set of labeled fragments by reproducing a template strand of DNA to be sequenced and stopping the replication process at one of the four bases by adding ddNTPs instead of dNTPs. There are four possible reaction mixes that result in A. T. G. or C, respectively (**Figures 5** and **6**).

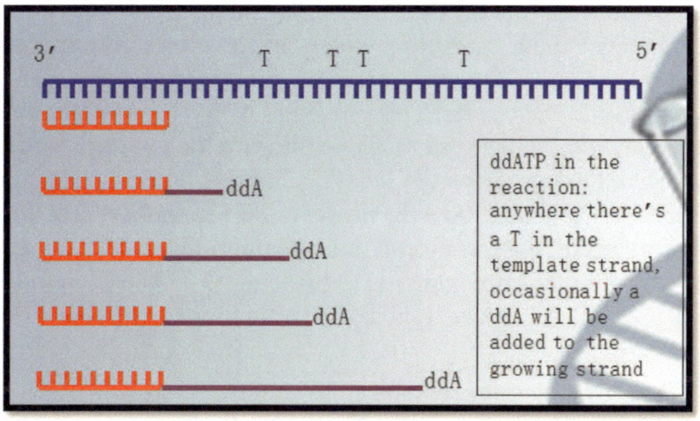

Figure 5.
Principle of reaction by addition of ddNTPs.

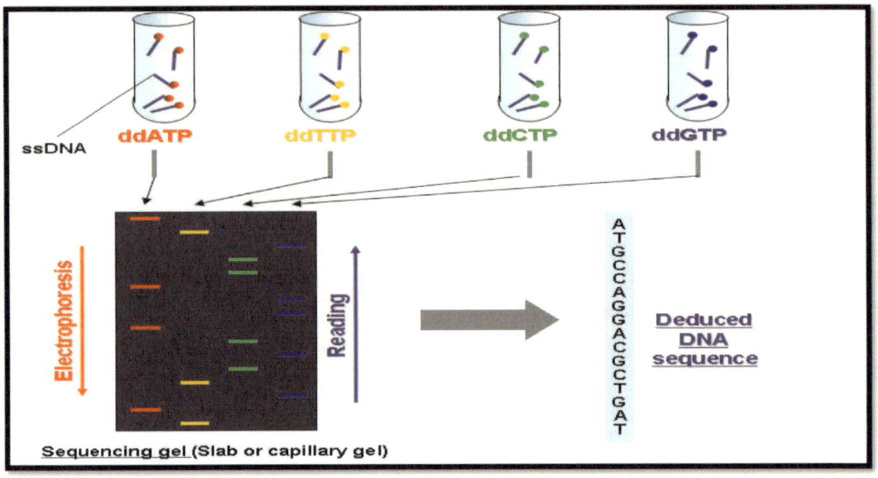

Figure 6.
Reading of sequences by Sanger's method.

3.2 2nd generation sequencing

3.2.1 Ion semiconductor sequencing

The Ion Torrent method relies on observing hydrogen ion release occurring as new nucleotides are incorporated into the expanding DNA template. A hydrogen ion is produced as a by-product in nature when a polymerase adds a nucleotide to a strand of DNA. Ion Torrent's Ion Personal Genome Machine (PGMTM) sequencer performs nucleotide incorporation in a highly parallel fashion using a high-density array of micro-machined wells. Each holds a paraphrased template. An ion-sensitive layer sits below the wells, followed by a specialized ion-sensor. An ion sensor picks up the ion because it alters the pH of the solution. The output voltage doubles, and the chip records two identical bases without scanning, camera, or light if there are two identical bases on the DNA strand. Ion Torrent technology establishes a direct link between the chemical and digital events rather than relying on the detection of light as in 454 pyrosequencing. Ion-sequencing chips for semiconductor ions may detect hydrogen ions. Similar to other semiconductor chips used in electronic equipment, these ion semiconductor chips are created and manufactured. These are made by cutting wafer-shaped pieces of silicon from a boule. Then, using photolithography, the transistors and circuits are pattern-transferred and subsequently etched onto the wafers. This procedure is carried out 20 times or more, resulting in a multilayer circuitry system. Depending on the ion semiconductor sequencing chip type utilized, the Ion Torrent PGM (Personal Genome Machine) produces a total data output of between 10 and 1000 MB. However, in September 2012, Ion Torrent launched its bigger system, the Ion Proton. It uses larger, higher-density chips, making it appropriate for whole- and exome-genome sequencing [3]. Even though Ion Proton can produce outputs up to 10 GB in size, it is significantly more expensive.

3.2.2 Illumina (Solexa) sequencing technology

Sequencing technology from Solexa Sequencing templates are immobilized on a custom flow cell surface created to present the DNA in a way that makes it accessible to enzymes while also assuring excellent surface stability and little nonspecific binding of fluorescently labeled nucleotides. By using solid-phase amplification, each molecule can be duplicated up to 1000 times nearby (diameter of one micron or less). Solexa sequencing technology may produce up to 10 million single molecule clusters per square centimeter densities without photolithography, mechanical spotting, or placing beads into wells. Sequencing-by-Synthesis The millions of clusters found on the surface of the flow cell are sequenced using Solexa sequencing using four unique fluorescently labeled modified nucleotides. Each cycle of the sequencing reaction can occur concurrently in the presence of all four of these nucleotides because of these nucleotides' uniquely created ability to have a reversible termination capability (A, C, T, and G). Because all four potential bases naturally compete, the polymerase can choose the suitable base to incorporate in each cycle more accurately than when only one nucleotide is present in the reaction mixture at a time. Homopolymers, for example, are sequences in which a specific base is repeated repeatedly. These sequences are addressed precisely and similarly to other sequences (**Figure 7**).

Analytical Process: A huge quantity of short sequence reads form the foundation of the Solexa sequencing method. To produce a consensus and assure high confidence in the identification of genetic differences, deep sampling with more than ten-fold even coverage is necessary. Sequence reads are compared to a reference to find such discrepancies. Deep sampling enables the use of weighted "majority voting" and statistical analysis, similar to traditional approaches, to discern between sequencing errors and homozygotes and heterozygotes. Each raw read base is given a quality

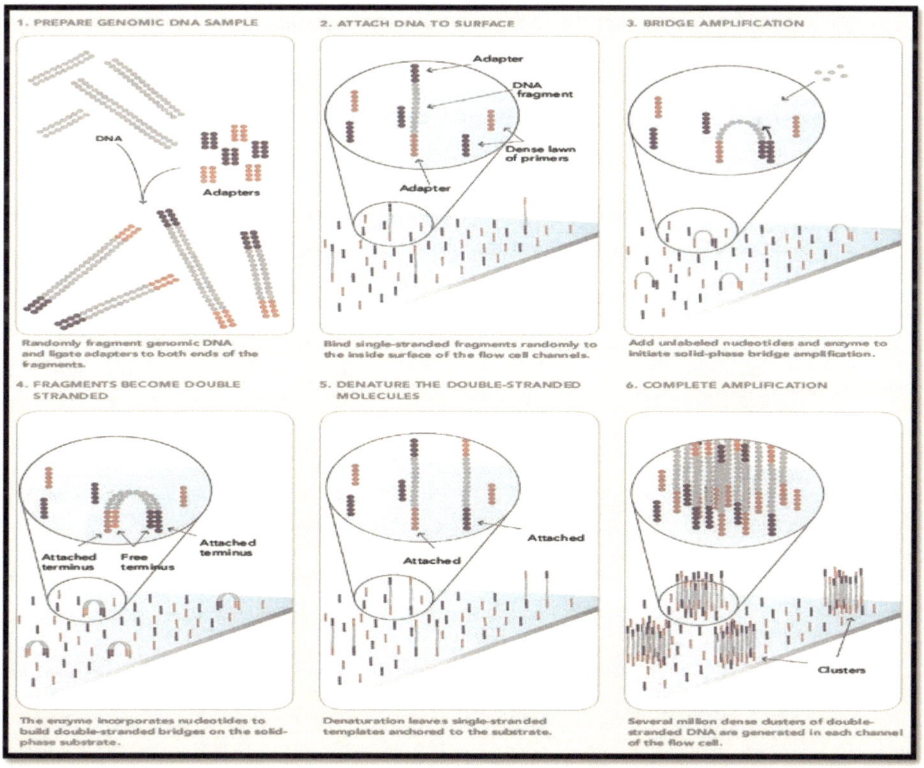

Figure 7.
Steps generated by Illumina sequencing - [4].

score, which the software uses to generate confidence scores and apply a weighting factor to calling differences.

3.2.3 Sequencing by ligation (SOLiD)

SOLiD is an enzymatic sequencing technique that uses DNA ligase, a biotechnological enzyme widely employed for its capacity to ligate double-stranded DNA strands (**Figure 8**). A ssDNA primer-binding region (sometimes referred to as an adapter) that has been conjugated to the target sequence (i.e., the sequence to be sequenced) on a bead is immobilized or amplified using emulsion PCR. A high bead density is then created by depositing these beads onto a glass surface, increasing the throughput of the method. After attaching beads, a primer of length N binds to an adaptor. The beads encounter an array of 8-mer probes in a library, each ending in a hydroxyl group and featuring various fluorescent dyes at the 5′ end. Bases 3–5 are variable, and bases 6–8 are inosine, while bases 1 and 2 complement the target nucleotides. Only a matching probe can attach to the primer-adjacent sequence. The 8-mer probe is then connected to the primer using DNA ligase. Using silver ions, the fluorescent dye is cleaved from the fragment through a phosphorothioate bond between bases 5 and 6, creating a 5′-phosphate group for fluorescence monitoring. Four different dyes, each with unique emission spectra, aid in detection. After the initial sequencing round, the extension product is dissociated, and a primer of length N-1 is employed for subsequent rounds. This iterative process continues with increasingly shorter primers (e.g., N-2, N-3, etc.), with fluorescence continuously monitored.

Known for its high accuracy (99.999% with a sixth primer), the SOLiD method is cost-effective due to its two-base sequencing approach (each base is read twice). Despite generating up to 30 Gb of data in a week-long run, its short read lengths limit its utility in certain applications, presenting a significant drawback.

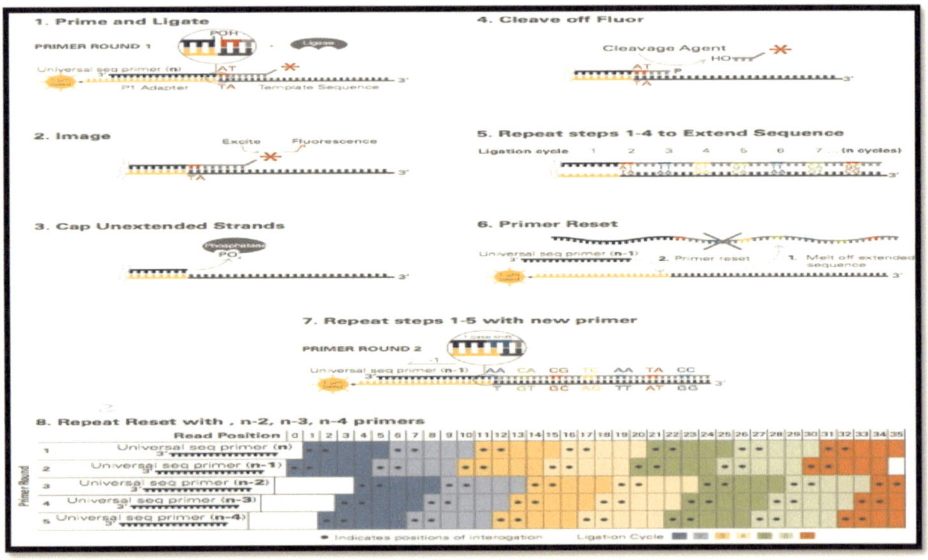

Figure 8.
Stepwise details of SOLiD method - [5].

DOI: http://dx.doi.org/10.5772/intechopen.1006294

3.2.4 454 pyrosequencing

The complementary strand is first removed after creating the polonies needed for sequencing using the emulsion PCR method. The four separate dNTPs are then progressively made to flow in and out of the wells across the polonies after the ssDNA sequencing primer hybridizes to the end of the strand (primer-binding region). Pyrophosphate is released when the proper dNTP is enzyme-incorporated into the strand. The pyrophosphate is changed into ATP when ATP sulfurylase and adenosine are present. With the aid of an ATP molecule, luciferin is converted to oxyluciferin, which emits light that a camera may capture. The quantity of base added determines the relative intensity of light. (i.e., a peak of twice the intensity indicates two identical bases have been added in succession) (**Figure 9**). One of the early triumphs of next-generation sequencing was pyrosequencing, which 454 Life Sciences created; in fact, 454 Life Sciences created the first next-generation sequencer that was readily available for purchase. Rival technologies overshadowed the technique, and in 2013, 454 Life Sciences and the 454 pyrosequencing platform were shut down by new owners, Roche.

3.2.5 Polony sequencing

It is a cheap but exact multiplex sequencing method that simultaneously reads millions of immobilized DNA sequences. Dr. George Church invented this method first at Harvard Medical College. It sequenced the entire *Escherichia coli* genome using an *in vitro* paired-tag library, emulsion PCR, an automated microscope, and ligation-based sequencing chemistry at a cost roughly ten times lower than Sanger sequencing.

3.3 Sequencing epigenetic modifications

It has lately been evident that the genetic code does not carry all the information needed by organisms, just as Next-generation sequencing allowed for large genomic sequencing. DNA base epigenetic changes, particularly those to 5-methylcytosine, also provide significant information. All second-generation sequencing technologies rely on PCR, just like Sanger sequencing. Hence, they are unable to sequence changed

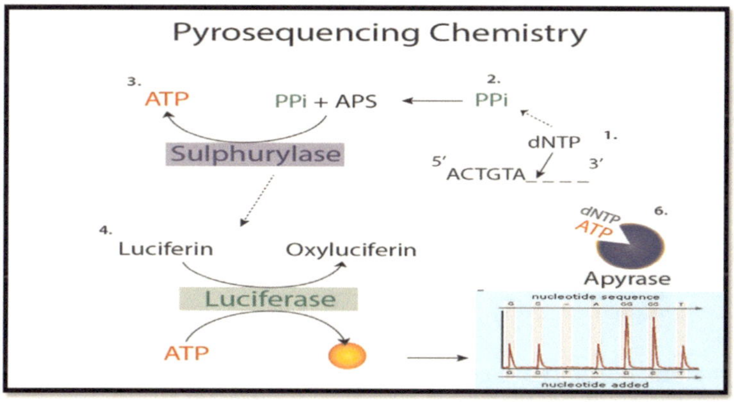

Figure 9.
Flowchart of pyrosequencing.

DNA bases. The PCR enzymes regard both 5-methylcytosine and 5-hydroxymethylcytosine as cytosine; as a result, epigenetic information is lost during sequencing.

3.3.1 Bisulfite sequencing

Bisulfite sequencing takes use of the difference between cytosine and 5-methylcytosine's reactivity to bisulfite: cytosine is deaminated by bisulfite to generate uracil, which, when sequenced, reads as T. (i.e., reads as C). When two sequencing runs are carried out simultaneously, one with and one without bisulfite treatment, the variations between the two runs' outputs point to methylated cytosines in the original sequence. Since the strands are no longer complimentary after exposure to bisulfite, this method can also be applied to dsDNA. Another significant epigenetic change, 5-hydroxymethylcytosine, interacts with bisulfite to produce cytosine-5-methylsulfonate (which reads as C when sequenced). This makes things a little more complicated and prevents the use of bisulfite sequencing as a reliable methylation indicator in and of itself.

3.3.2 Oxidative bisulfite sequencing

Oxidative bisulfite sequencing includes a chemical oxidation phase that uses potassium perruthenate, KRuO4, to change 5-hydroxymethylcytosine into 5-formylcytosine prior to bisulfite treatment. 5-Formylcytosine is deformylated and deaminated to form uracil by bisulfite treatment. Now, to distinguish between cytosine, 5-methylcytosine, and 5-hydroxymethylcytosine, three independent sequencing runs are required.

3.4 3rd generation sequencing/high throughput sequencing

The huge demand for low-cost sequencing sparked the invention of high-throughput sequencing technologies, which parallelize the sequencing process and produce thousands or millions of genomes at once. Researchers may sequence target regions in-depth, swiftly sequence entire genomes, detect splice sites and RNA sites in the transcriptome, identify epigenetic modifications, study microbial diversity and evolution, and sequence diseased samples to find differences using NGS technology (which may be disease biomarkers or therapeutic targets) [6]. The declining cost and increasing precision of NGS have made it an essential tool for genetic research and have been used in routine clinical practice for precise medicine. NGS is fundamentally constrained by the necessary infrastructure, which includes the computing power and storage needed as well as the expertise of the employees needed to analyze and comprehend the massive amounts of data produced by the system. Despite the rapid advancement of third-generation sequencing technologies, NGS offers incomparable benefits and would not be supplanted entirely for a while (including PacBio SMRT and Nanopore technologies).

3.4.1 Lynx therapeutics' massively parallel signature sequencing (MPSS)

MPSS, the first of the "next-generation" sequencing technologies, was created at Lynx Therapeutics in the 1990s. Sydney Brenner and Sam Eletr launched Lynx Therapeutics in 1992. MPSS is a very high throughput sequencing technology. It reveals practically every transcript in the sample and accurately indicates its expression level when applied to an expression profile. The MPSS bead-based method was sensitive to sequence-specific bias or loss of particular sequences since it employed a complicated strategy of adapter ligation followed by adapter decoding and reading the sequence in increments of four nucleotides. Yet the fundamental characteristics of the MPSS output, which included thousands of short DNA sequences, were typical of later "next-gen" data formats. For MPSS, these were typically applied to the sequencing of cDNA to quantify the gene expression levels. In 2004, Lynx Therapeutics merged with Solexa, and Illumina later acquired this business.

3.4.2 Helicos/Heliscope

The Cambridge, Massachusetts, USA-based Helicos Biosciences may be the first business to provide a "next-next" generation DNA sequencing system for single molecules [7]. According to Steve Lombardi, president and COO of Helicos, the single-molecule sequencing technology that Helicos plans to introduce in 2008 may study DNA or RNA, look at sequence variation, and possibly look at epigenetics. Our objective was to create a genetic analyzer capable of large-scale production. The HeliScope single-molecule sequencer is a device that uses Helicos' technology, also known as accurate single-molecule sequencing (tSMS), which is a sequencing-by-synthesis method for single molecules. DNA must be broken up into fragments of 100–200 base pairs in order to produce a sample. DNA fragments' ends must then be capped with adaptors of known sequence, frequently poly(A) tails. Harris claims that by covalently attaching a poly(T) primer to a surface, "We can catch that poly(A) tail." The distance between the collected fragments is crucial for imaging during tSMS. According to Harris, the distance between captured molecules must be at least as great as the resolution of optical microscopy. A single tagged nucleotide and polymerase mixture is then applied across the surface once the fragments have been joined. All caught fragments with the complementary nucleotide in the first free position have the designated nucleotides incorporated by the polymerase. The HeliScope camera photographs the entire surface after the incorporation event and identifies all acquired fragments with an incorporated tagged nucleotide. The following stage of the procedure, where the labeling dye is broken off to allow the polymerase to incorporate another labeled nucleotide into the fragment, is one of the keys to making tSMS function. According to Harris, their instrument is now able to get precise reads on collected fragments ranging from 25 to 45 bases over billions of strands by repeatedly cycling through this procedure using each of the four nucleotides. However, a unique aspect of this platform is that no clonal amplification is required [8]. Another difficulty for the Helicos developers was imaging during each cycle. A huge number of data points are produced throughout the sequencing process as a result of the capture of so many high-resolution pictures. Harris claims that even while each run of the Heliscope requires 14 terabytes of computer storage, those 14 terabytes can store a tremendous quantity of sequence data; the imaging system can deliver one gigabase of sequence each hour. Harris and Lombardi are keen to point out that this allows the system a ton of headroom for future advancements in sequencing chemistry, even if the current version is not even close to this goal right now.

3.4.3 Pacific bioscience- single molecule SMRT(TM) sequencing

PacBio sequencing is a real-time sequencing technique that does not call for a break in between read processes. PacBio sequencing is referred to as third-generation sequencing because of these characteristics that set it apart from SGS (TGS). Here, we will provide a summary of PacBio sequencing's workings and mechanisms. In comparison to SGS techniques, PacBio sequencing provides substantially longer read lengths and faster runs. Still, it is constrained by lesser throughput, a higher error rate, and a higher cost per base. We will look into hybrid sequencing approaches that combine SGS and PacBio sequencing to overcome each technology's drawbacks separately because each technology's benefits are complementary. The applications of PacBio sequencing to various research areas include genome, transcriptome, and epigenetics. Due to new computational methods and advancements in sequencing technology, PacBio sequencing can now be used to study much larger genomes, including the human genome. In contrast, its reasonable applications to genomics research were initially restricted to the completion of relatively small microbial genomes. Because of its lengthy read lengths, PacBio sequencing can identify and quantify isoforms, including novel ones, especially when combined with SGS [9]. Moreover, PacBio sequencing kinetic enables the direct identification of base modifications like N6-methyladenine (m6A) and N4-methylcytosine by tracking the intervals between base incorporations (m4C). During the target DNA molecule's replication phase, PacBio sequencing records sequence data. The target double-stranded DNA (dsDNA) molecule is transformed into the template, known as an SMRTbell, which is a closed, single-stranded circular DNA molecule. When an SMRT cell, a type of chip, is loaded with a sample of SMRTbell. A SMRTbell diffuses into a zero-mode waveguide (ZMW), a sequencing component offering the tiniest light detection space. Each ZMW has a single immobilized polymerase at the bottom that can attach to any of the SMRTbell's hairpin adaptors and initiate replication. The SMRT cell is supplemented with four fluorescently labeled nucleotides that produce different emission spectra. A light pulse that is created while the polymerase retains a base identifies the base. A "movie" of light pulses that corresponds to each ZMW of an SMRT cell records the replication activities in each ZMW, and the pulses can be viewed as a series of bases (called a continuous long read, CLR). The newest platform, PacBio RS II, typically generates sequencing videos that are between 0.5 and 4 hours long. After the polymerase repeats one strand of the target dsDNA, adapter bases can be added before replicating the second strand because the SMRTbell forms a closed circle. In a single CLR, both strands can be sequenced numerous times (referred to as "passes") if the polymerase has a long enough half-life. In this case, the CLR can be divided into numerous reads (referred to as subreads) by identifying and removing the adaptor sequences. A circular consensus sequence (CCS) read with greater precision is produced by the consensus sequence of several subreads in a single ZMW. A CCS read cannot be produced if the target DNA is too lengthy to be sequenced several times in a CLR; instead, only a single subread is output. Because PacBio sequencing occurs in real-time, it is possible to study kinetic variation interpreted from the light-pulse movie to find base changes like methylation.

3.4.4 Oxford nanopore

ONT employs a unique method in which native DNA molecules are pulled through tiny pores (nanopores) that only allow one DNA molecule at a time, in contrast to PacBio, which is a "sequencing by synthesis" platform [10]. Sensors track variations in the ionic current caused by each nucleotide as it passes through the pore to determine how the DNA is moving. This data provides the signal needed for base-calling and can be seen as a "squiggle plot." Theoretically, sequencing continues until the DNA fragment's end or the pore is physically blocked, enabling hitherto unheard-of read lengths that could greatly enhance *de novo* genome assembly and the discovery of structural changes in huge genomes. This is especially important in plant genomes containing highly repetitive regions derived from transposons and tandem repeats.

4. Applications

- *Understanding the function of specific sequences and disease-related sequences*: Gene sequencing allows researchers to determine the exact sequence of nucleotides in a particular gene or genomic region (**Table 2**). This knowledge is crucial for understanding how genes function and how mutations or variations in these sequences can lead to diseases. By comparing sequences from healthy individuals with those affected by diseases, researchers can pinpoint specific mutations or variants that may be responsible for causing or predisposing individuals to certain diseases. This information is vital for developing targeted therapies, genetic counseling, and personalized medicine approaches.

Year	Researchers/Group	Gene Sequencing Technique	Trait/Quality Improved	Type of Organism	References
2001	International Rice Genome Sequencing Project (IRGSP)	Whole Genome Sequencing	Yield, Disease Resistance	Rice	Goff et al. [11]
2008	Wellcome Trust Sanger Institute	Next-Generation Sequencing (NGS)	Disease Resistance, Growth	Wheat	Brenchley et al. [12]
2010	J. Craig Venter Institute	Whole Genome Sequencing	Microbial Community Profiling	Marine Microbes	Rusch et al. [13]
2012	Human Microbiome Project Consortium	16S rRNA Sequencing	Microbial Community Characterization	Humans	Human Microbiome Project Consortium [14]
2014	International Wheat Genome Sequencing Consortium (IWGSC)	Whole Genome Sequencing	Yield, Disease Resistance	Wheat	International Wheat Genome Sequencing Consortium (IWGSC) [15]
2015	University of California, Davis	Whole Genome Sequencing	Flavor, Nutritional Content	Tomato	Tomato Genome Consortium [16]
2016	Beijing Genomics Institute (BGI)	Whole Genome Sequencing	Salt Tolerance	Rice	Zhang et al. [17]
2017	University of Queensland	RNA Sequencing	Drought Tolerance	Sorghum	Fracasso et al. [18]
2017	Cold Spring Harbor Laboratory	CRISPR-Cas9 Genome Editing	Herbicide Resistance	Cotton	Li et al. [19]
2018	Broad Institute	CRISPR-Cas9 Genome Editing	Disease Resistance, Yield	Maize	Shi et al. [20]

Year	Researchers/Group	Gene Sequencing Technique	Trait/Quality Improved	Type of Organism	References
2018	Japan Tobacco Inc.	Whole Genome Sequencing	Aroma, Flavor	Tobacco	Sierro et al. [21]
2019	Chinese Academy of Sciences	Whole Genome Sequencing	Disease Resistance	Wheat	Qiao et al. [22]
2019	University of Illinois at Urbana-Champaign	CRISPR-Cas9 Genome Editing	Nutrient Efficiency	Maize	Shi et al. [23]
2019	Broad Institute	Single-Cell RNA Sequencing	Immune Response Profiling	Mice	Zheng et al. [24]
2020	Chinese Academy of Agricultural Sciences	RNA Sequencing	Growth Rate, Stress Resistance	Soybean	Fang et al. [25]
2020	Cornell University	Genotyping-by-Sequencing (GBS)	Pest Resistance	Potato	Hamilton et al. [26]
2020	University of California, Riverside	CRISPR-Cas9 Genome Editing	Water Use Efficiency	Tomato	Yu et al. [27]
2021	National Institutes of Health (NIH)	Whole Exome Sequencing	Disease Mutation Identification	Humans	Bamshad et al. [28]
2021	John Innes Centre	RNA Sequencing	Heat Tolerance	Barley	Thirumalaikumar et al. [29]
2021	European Bioinformatics Institute	Single-Cell RNA Sequencing	Immune Cell Profiling	Humans	Villani et al. [30]
2021	Wageningen University & Research	Whole Genome Sequencing	Cold Tolerance	Cucumber	Bo et al. [31]
2022	Australian National University	Whole Genome Sequencing	Nitrogen Use Efficiency	Wheat	Good and Beatty [32]
2023	University of São Paulo	Genotyping-by-Sequencing (GBS)	Nutritional Quality	Maize	Silva et al. [33]
2023	National Institute of Agricultural Botany (NIAB)	Whole Genome Sequencing	Disease Resistance	Potato	Tiwari et al. [34]

Table 2.
List of researchers involved in the improvement in traits/quality improvement in animal and plant species using various gene sequencing techniques.

- *Detection of mutations through comparative DNA sequence studies*: Comparative DNA sequence studies involve analyzing the sequence of DNA from different individuals, species, or populations to identify variations and mutations. These mutations can include single nucleotide polymorphisms (SNPs), insertions, deletions, and structural variations. Identifying mutations is essential for understanding genetic diversity, evolutionary relationships, and disease susceptibility across populations. In clinical settings, detecting mutations through sequencing helps diagnose genetic disorders, predict disease risk, and tailor treatments based on an individual's genetic profile.

- *DNA fingerprinting*: DNA fingerprinting, also known as DNA profiling or genetic fingerprinting, is a technique used to identify and distinguish between individuals based on their unique DNA sequences. It involves analyzing specific regions of an individual's genome that contain highly variable sequences, such as short tandem repeats (STRs) or variable number tandem repeats (VNTRs). DNA fingerprinting is widely used in forensic science to identify suspects or victims in criminal investigations, establish paternity or familial relationships, and resolve immigration disputes. It is also employed in conservation biology to study genetic diversity within and between species.

- *Completion of the Human Genome Project (HGP)*: The Human Genome Project was an international scientific effort that aimed to map and sequence the entire human genome, which comprises approximately 3 billion base pairs of DNA. Completed in 2003, the HGP provided a comprehensive reference map of the human genome, identifying all genes and their locations on chromosomes. This monumental achievement has had far-reaching implications for biomedical research, medicine, and genetics. It laid the foundation for subsequent large-scale genomic studies, enabled the discovery of disease-related genes, facilitated the development of genomic technologies, and promoted advances in personalized medicine.

- *Forensics*: DNA sequencing is employed in forensic science to identify individuals based on their unique DNA sequences. It is instrumental in criminal investigations, where biological evidence such as hair, nails, skin, or blood samples recovered from crime scenes can be sequenced to establish links to suspects or victims.

- *Medicine*: In medical research, DNA sequencing plays a crucial role in identifying genes associated with hereditary or acquired diseases. Scientists utilize genetic engineering techniques, including gene therapy, to pinpoint defective genes and develop treatments aimed at replacing them with healthy alternatives. This approach holds promise for personalized medicine, tailoring therapies to individuals based on their genetic makeup.

4.1 Human genome project

The Human Genome Project, launched on October 1, 1990, was a monumental international effort involving scientists from 20 institutions across six nations.

It aimed to map and sequence the entire human genome, which comprises approximately 3 billion chemical nucleotide bases (A, C, T, and G). This endeavor revealed fascinating insights, such as the fact that in 99.9% of all individuals, these nucleotide bases are identical. The largest known human gene, dystrophin, consists of 2.4 million bases, although the average gene size is around 3000 bases. Initially expected to contain between 80,000 and 140,000 genes, researchers discovered that the human genome likely contains between 30,000 and 80,000 genes, with over 50% lacking well-defined functions. By June 26, 2001, a working draft of the human genome was publicly released, marking a significant milestone. This project, however, came with a staggering cost of $3 billion. The sequencing efforts were concentrated at five major locations across the United States and the United Kingdom, collectively known as the G5. These included the Broad Institute/Whitehead Institute for Biomedical Research at MIT in Cambridge, Massachusetts; Washington University in St. Louis; Baylor College of Medicine in Houston; the Joint Genome Institute of the Department of Energy in Walnut Creek, USA; and the Wellcome Trust Sanger Institute in Cambridge, UK (formerly the Sanger Centre), which focused on specific chromosomes (1, 6, 10, 11, 13, 20, 22, and X). The collaborative nature of the Human Genome Project involved countries such as the USA, China, Germany, Japan, UK, and France, underscoring its global significance in advancing our understanding of human genetics and laying the foundation for subsequent genomic research and medical breakthroughs.

- Agriculture: DNA sequencing has played a vital role in agriculture. The mapping and sequencing of the whole genome of microorganisms has allowed agriculturists to make them useful for crops and food plants.

4.2 Plant genome projects

Recent advancements in DNA sequencing technologies have revolutionized plant genome projects by significantly reducing the cost and time required to sequence entire genomes [35]. The primary objective of these projects is to achieve whole genome sequencing, which serves as the cornerstone of genomic research. By focusing on model organisms with small genomes or those suitable for genetic studies, early efforts in plant genomics aimed to identify every gene within species and unravel their functions. One of the pioneering endeavors began in 1990 under the leadership of the National Science Foundation (NSF), spearheading an international, multiagency initiative to comprehensively catalog every gene in the *Arabidopsis thaliana* genus by 2000. *Arabidopsis thaliana*, a flowering plant, emerged as the first plant species to have its genome fully sequenced, marking a significant milestone in botanical research. This plant species has since become invaluable as a model organism for gene discovery and understanding gene functions due to its genetic tractability and evolutionary significance. In 2008, the Genome Initiative introduced the 1001 Genomes Project, a collaborative effort involving eleven institutions worldwide. This ambitious project aimed to sequence the genomes of 1001 Arabidopsis strains collected from diverse geographical locations. By analyzing this extensive genetic diversity, researchers sought to uncover the genetic variations underlying plant adaptation to different environments. These initiatives highlight the global collaboration and technological advancements driving plant genomics forward, offering insights crucial for agriculture, environmental sustainability, and biotechnological applications.

4.3 1000 Genome project

The 1000 Genome Project, initiated in 2008 under the leadership of Drs. Gane Ka-Shu Wong and Michael Deyholos from the University of Alberta, aims to identify variations in whole genome sequences across 1000 different plant species. Over the years, the project has focused on compiling the transcriptomes—expressed genes—of these species, revealing hundreds of genes that are either absent in some strains or present in extra copies due to millions of small genetic changes. This genetic flexibility enables plants to exhibit a diverse array of molecular gene products and adaptability. To achieve rapid and precise sequencing, the project utilizes 28 next-generation Illumina Genome Analyzer DNA sequencing machines at the Beijing Genomics Institute (BGI - Shenzhen, China). Each machine can process up to 3 billion base pairs per run, facilitating efficient analysis of plant samples. The initiative is set to expand the number of plant species with extensively sequenced genomes nearly a hundredfold, representing a significant leap forward in botanical research and biotechnology applications.

This global effort involves a partnership that has nearly finalized the selection of plant species to be sequenced, prioritizing those known for their functional biochemical properties. Additionally, other species have been chosen to fill knowledge gaps and elucidate unresolved evolutionary relationships within the plant kingdom. The 1000 Genome Project promises to deepen our understanding of plant genetics, offering insights crucial for agricultural innovation, conservation efforts, and environmental sustainability.

4.4 *Arabidopsis* genome project

The Arabidopsis Genome Project, initiated by American scientists Meyerowitz, Somerville, and Goodman, marked a pioneering effort in plant genomics. Beginning with the production of the initial RFLP map by Meyerowitz, followed by genomic comparisons conducted by the Goodman Group, the project aimed to sequence and understand the entire genome of *Arabidopsis thaliana*. By 1998, physical maps of each chromosome were completed, with chromosomes 1, 3, and 5 fully sequenced and analyzed by 2000. The formal conclusion of the *Arabidopsis* Genome Sequencing Project in late 2000 marked a significant milestone in plant genetics. *Arabidopsis thaliana*, with a genome approximately 146 million base pairs long, features genes spaced every 4–5 kilobases, encoding between 20,000 and 50,000 genes across various functional classes. These genes include approximately 19,117 proteins with known transcripts, some of which carry signal sequences directing their products to organelles like mitochondria or chloroplasts. The genome exhibits extensive DNA repetition in telomeric and centromeric regions, where pseudogenes and transposons are prevalent. Additionally, gene and DNA duplications between chromosomes contribute to the plant's genetic complexity. Chromosome 4 of *Arabidopsis* is particularly noteworthy as it houses the entire mitochondrial genome, highlighting the species' intricate genetic architecture. The comprehensive insights gained from sequencing *Arabidopsis thaliana* have profoundly influenced plant biology, serving as a crucial model for understanding gene functions, evolutionary processes, and genetic diversity in plants.

4.5 Rice genome project

Closely related to other cereals but with a significantly smaller genome than those of those plants, it is the first agricultural plant whose entire genome has been sequenced. The International Rice Genome Sequencing Project (IRGSP) was launched in September 1997 during a workshop held in conjunction with the International Symposium on Plant Molecular Biology in Singapore. Scientists from Japan initiated the Rice Genome Programme (RGP) in 1991. At the event, researchers from several countries decided to work together internationally to sequence the rice genome.

As a result, 6 months later, representatives from Japan, Korea, China, the United Kingdom, and the United States gathered to set the rules at Tsukuba. The participants consented to the exchange of information and the prompt uploading of annotated DNA sequences and physical maps to public databases. The IRGSP Working Group, comprising representatives from each participating country, develops IRGSP regulations and finishing standards. The IRGSP has expanded to include 11 nations (**Figure 10**).

The International Rice Genome Sequencing Project (IRGSP) undertook the sequencing of *Oryza sativa* ssp. *japonica* cv. Nipponbare using a hierarchical clone-by-clone method, employing bacterial (BACs) and P1 artificial chromosome clones (PACs). This initiative began in 1997 and was followed by the RGP's second phase in 1998. Monsanto contributed significantly in 2000 by generating a draft genome of the *Japonica* rice variety, estimating its DNA content to range between 420 and 466 million base pairs from 3391 BAC clones. Subsequently, the Beijing Genomic Institute and Syngenta published draft sequences of Indian and other Japonica rice varieties in 2002 and 2005, respectively. Gen Bank data on rice indicated that 28,282,731 bases of sequences from rice EST sequencing initiatives have been contributed. Comparative genomic analyses have shown that the number of genes in rice is similar to *Arabidopsis*, with approximately 40,000 genes (37,344 coding genes) identified,

Rice Sequencing Participants and Chromosome Assignments

Site	Chromosome
Rice Genome Research Program (RGP; Japan)	1,6,7,8
Korea Rice Genome Research Program (Korea)	1
CCW (United States) CUGI (Clemson University) Cold Spring Harbor Laboratory Washington University Genome Sequencing Center	3,10
TIGR (United States)	3,10
PGIR (United States)	10
University of Wisconsin (United States)	11
National Center for Gene Research, Chinese Academy of Sciences (China)	4
Indian Rice Genome Program (University of Delhi)	11
Academia Sinica Plant Genome Center (Taiwan)	5
Genoscope (France)	12
Universidad Federal de Pelotas (Brazil)	12
Kasetsart University (Thailand)	9
McGill University (Canada)	9
John Innes Centre (United Kingdom)	2

Figure 10.
Rice genome sequencing partners.

including 2859 that are specific to rice and other cereals. The gene density in rice averages one gene per 9.9 kilobases, which is less dense compared to Arabidopsis. The sequenced segments cover 95% of the rice genome and 99% of its euchromatin, with the average rice gene spanning 2.2 kilobases and containing 3.9 exons and 2.9 introns. This equates to one gene approximately every 5.7 kilobases. Rice is notable for its rich diversity of transposon superfamilies, comprising at least 35% of known transposons. Additionally, organellar DNA fragments constitute 0.38 to 0.43% of the nuclear genome, indicating ongoing transfers of organellar DNA. These findings underscore rice's genetic complexity and its significance in agricultural research and food security globally.

4.6 Wheat genome project

In a significant development, a team of over 200 scientists from 73 institutions in 20 countries, including 18 Indian scientists from the ICAR-National Research Centre on Plant Biotechnology in New Delhi, Punjab Agricultural University in Ludhiana, and Delhi University South Campus, has unraveled the intricate genome of the bread wheat research variety "Chinese Spring," a challenge that was previously regarded as impossible. The bread wheat genome, a hexaploid with a highly intricate structure combining three sets of chromosomes (AABBDD), is notably 40 times larger than the rice genome and five times larger than the human genome. Over a span of 13 years and costing approximately US $75 million, the International Wheat Genome Sequencing Consortium (IWGSC) achieved a major milestone by publishing the comprehensive DNA sequence in the esteemed journal "Science" on August 17, 2018. The reference genome of bread wheat, derived from the Chinese Spring variety, provides detailed insights into its genetic makeup. It predicts a total of 107,981 protein-coding genes and achieves a genome coverage of 94%, amounting to approximately 145,000 lakh base pairs across the wheat genome. Notably, the sequencing task for chromosome 2A, which constitutes about 5% of the wheat genome, was entrusted to the Indian scientific community. The genome of modern bread wheat (*T. aestivum*) with 42 chromosomes spans 17 Gb and comprises over 80% repetitive sequences with only 2% coding sequence. The sub-genomes A, B, and D exhibit high sequence similarity and collectively harbor 123,201 gene loci that are uniformly dispersed.

5. Advantages

- *Forensic Identification*: DNA testing has revolutionized forensic science by providing highly accurate identification. Unlike fingerprints, DNA profiles are unique to individuals, making it a powerful tool in criminal investigations. By analyzing DNA evidence from blood, skin, or hair found at crime scenes, forensic scientists can match it to suspects or establish links between individuals, providing critical evidence for solving crimes.

- *Enhanced Crime Solving*: Advances in DNA technology have enabled the re-examination of old crime scene evidence with improved testing methods. This has led to the resolution of cold cases that were previously unsolved, showcasing the retrospective power of DNA analysis in criminal justice.

- *Medical Applications*: DNA testing plays a crucial role in medical genetics, allowing for the screening of genetic diseases and identification of genetic risk factors. This is particularly valuable in prenatal testing, where prospective parents can assess the genetic health of embryos before implantation, aiding in informed decision-making during fertility treatments.

- *Disaster Victim Identification*: In mass casualty incidents such as plane crashes or terrorist attacks, DNA testing facilitates rapid and accurate identification of victims when traditional methods like dental records or fingerprints are unavailable or insufficient. This ensures timely repatriation of remains to families and proper closure in the aftermath of disasters.

- *Unchanging Identity Marker*: DNA provides a stable and unalterable identifier that remains consistent across all tissues and organs of an individual's body. This permanence makes DNA fingerprinting a reliable method for identification even when other physical identifiers may be compromised or unavailable.

- *High Accuracy*: The probability of a DNA match between unrelated individuals is extremely low, typically ranging from 1 in 7000 to 1 in a billion, depending on the uniqueness of genetic markers being compared. This high degree of accuracy makes DNA testing more reliable than eyewitness testimony in legal proceedings.

- *Advances in Personalized Medicine*: Gene sequencing also contributes to personalized medicine by identifying genetic variations that predispose individuals to certain diseases or influence their response to medications. This allows healthcare providers to tailor treatment plans and preventive measures based on individual genetic profiles, improving patient outcomes.

Gene sequencing has revolutionized agricultural sciences by offering several key advantages, each contributing to the improvement of crop varieties, sustainable farming practices, and global food security:

- *Precision Breeding*: Gene sequencing enables the precise identification of genes associated with desirable traits in crops. For example, researchers have used gene sequencing to identify genes responsible for disease resistance in wheat. By pinpointing these genes, breeders can develop new varieties that are resistant to devastating diseases such as wheat rust, without relying on chemical treatments. This not only protects yields but also reduces environmental impact.

- *Accelerated Crop Improvement*: Traditional breeding methods rely on observing and selecting plants with desired traits over generations. Gene sequencing expedites this process by allowing breeders to identify genetic markers associated with traits of interest early in the breeding cycle. For instance, in maize breeding, genetic markers linked to drought tolerance have been identified through sequencing, enabling breeders to develop drought-resistant varieties more efficiently.

- *Enhanced Understanding of Genetic Diversity*: Gene sequencing provides insights into the genetic diversity of crop species. This knowledge is crucial for conserving genetic resources and exploiting wild relatives that may carry genes for traits such as heat tolerance or pest resistance. For example, in rice research, sequencing has revealed diverse genetic backgrounds in wild rice species that can be crossed with cultivated rice to introduce novel traits.

- *Disease and Pest Management*: By identifying genes associated with disease resistance, gene sequencing helps breeders develop crops that are naturally resilient to pathogens. In potatoes, for instance, sequencing has facilitated the development of varieties resistant to late blight, reducing the need for fungicides. Similarly, in cotton, sequencing has aided in breeding varieties resistant to bollworms, reducing pesticide use and environmental impact.

- *Nutritional Enhancement*: Gene sequencing plays a crucial role in biofortification efforts aimed at enhancing the nutritional quality of crops. For example, in maize and rice, sequencing has been used to identify genes responsible for the synthesis of essential vitamins and minerals, such as vitamin A and iron. This has led to the development of biofortified varieties that address specific nutrient deficiencies in populations dependent on these staple foods.

- *Climate Resilience*: Climate change poses challenges to agricultural productivity. Gene sequencing helps breeders develop climate-resilient crops by identifying genes associated with traits such as heat and drought tolerance. In wheat breeding, for example, sequencing has been instrumental in identifying genes linked to heat stress tolerance, enabling the development of varieties capable of maintaining yield under high temperatures.

- *Economic Benefits*: Developing improved crop varieties through gene sequencing offers economic benefits to farmers and agricultural industries. For instance, in soybeans, sequencing has been used to develop varieties with higher oil content, enhancing profitability for growers and processors. Similarly, in fruit crops like apples, sequencing has supported breeding efforts to develop varieties with improved fruit quality and shelf life, meeting consumer demand more effectively.

Gene sequencing plays a pivotal role in molecular breeding by offering several key advantages that enhance the efficiency and precision of developing new crop varieties:

- *Identification of Molecular Markers*: Gene sequencing enables the discovery of molecular markers linked to desirable traits such as disease resistance, drought tolerance, and yield potential. For example, in maize breeding, sequencing has identified genetic markers associated with resistance to maize lethal necrosis disease. These markers allow breeders to select plants with desired traits more efficiently, accelerating the breeding process.

- *Marker-Assisted Selection (MAS)*: With gene sequencing, breeders can employ MAS, where specific genetic markers identified through sequencing are used to select plants with target traits. This approach reduces the need for time-consuming and resource-intensive field trials by enabling early identification of promising candidates. In rice breeding, MAS has been used to develop varieties with improved cooking quality traits, meeting consumer preferences more effectively.

- *Genomic Selection*: Gene sequencing facilitates genomic selection, where genomic information is used to predict the breeding value of plants based on their genetic makeup. This approach leverages large-scale genomic data to enhance breeding accuracy and efficiency. In livestock breeding, genomic selection has revolutionized breeding programs by predicting traits such as milk production or disease resistance based on genomic markers identified through sequencing.

- *Understanding Complex Traits*: Gene sequencing provides insights into the genetic basis of complex traits, such as yield components and nutritional quality. For instance, in tomato breeding, sequencing has revealed genes responsible for fruit size and flavor profiles. This knowledge guides breeders in selecting parental lines for hybridization to achieve desired trait combinations efficiently.

- *Accelerated Trait Introgression*: With gene sequencing, breeders can precisely introgress genes from wild or exotic germplasm into cultivated varieties. For example, in wheat breeding, sequencing has facilitated the introgression of genes for rust resistance from wild relatives into commercial cultivars. This approach broadens the genetic base of cultivated crops, enhancing resilience to evolving pests and diseases.

- *Targeted Genome Editing*: Advances in gene sequencing have enabled targeted genome editing techniques such as CRISPR-Cas9. This technology allows precise modifications of specific genes to introduce beneficial traits or silence undesirable traits. Often, only a single nucleotide needs to be edited or many different loci need to be targeted simultaneously to make the required change in a phenotype; however, in each case, it may be necessary to sequence the genome to confirm that other changes have not occurred [36]. In soybean breeding, genome editing has been used to enhance oil content and improve resistance to soybean cyst nematode, a devastating pest.

- *Integration with Omics Technologies*: Gene sequencing integrates seamlessly with other omics technologies such as transcriptomics, proteomics, and metabolomics. This multi-omics approach provides comprehensive insights into the molecular mechanisms underlying complex traits and their interactions with environmental factors. In maize research, combined omics analyses have elucidated pathways involved in stress responses, guiding breeding efforts for resilience under changing climatic conditions.

Overall, gene sequencing has transformed various fields and empowered researchers and breeders with advanced tools to understand, manipulate, and utilize genetic information for the sustainable enhancement of crop productivity and efficiency of developing new crop varieties with improved traits, nutritional quality, and resilience to environmental challenges, from forensic science to medical diagnostics, disaster response, and personalized healthcare. These advancements are crucial for ensuring food security and sustainability in a changing global climate. The ability of sequencing to provide precise and actionable genetic information underscores its indispensable role in modern society.

6. Demerits

Gene sequencing, while transformative in various scientific fields, also presents several challenges and limitations:

7. Technical and methodological limitations

- *Cost and Time Intensiveness*: Gene sequencing can be prohibitively expensive and time-consuming, especially for whole genome sequencing breeding programs aiming to sequence multiple individuals or populations. The equipment and expertise required for sequencing, data analysis, and interpretation contribute significantly to the costs. Additionally, despite advancements, certain sequencing techniques still require considerable time to generate results and hence can limit accessibility for smaller research programs and developing countries. Also, the cost-effectiveness and scalability of gene sequencing for large-scale molecular breeding programs may pose practical challenges.

- *Data Management and Analysis Complexity*: The enormous volume of data generated from gene sequencing poses challenges in terms of storage, management, and analysis. Handling big data requires sophisticated computational tools and bioinformatics expertise, which may not be readily available in all research settings.

- *Accuracy and Errors*: Despite technological advancements, gene sequencing methods are not infallible. Errors can arise from various sources such as PCR amplification, sequencing chemistry, and bioinformatics algorithms. These errors can lead to inaccuracies in the final sequence data, impacting downstream analyses and conclusions.

- *Sample Quality and Availability*: The quality and availability of biological samples can affect the outcome of gene sequencing experiments. Samples may degrade over time or be contaminated, leading to compromised sequencing results.

- *Complexity of Traits*: Many agricultural traits of interest, such as yield, disease resistance, and stress tolerance, are controlled by multiple genes interacting with environmental factors. Identifying and sequencing all relevant genes for complex traits can be challenging and may not fully capture the genetic basis of trait variation.

- *Variability Across Genotypes*: Agricultural crops often exhibit significant genetic variability across different varieties and cultivars. Conducting comprehensive gene sequencing across diverse genetic backgrounds is resource-intensive and may not always provide universally applicable insights.

- *Integration into Breeding Programs*: Integrating gene sequencing data into conventional breeding programs requires validating genetic markers and understanding their functional relevance. Incorporating genomic information effectively into breeding strategies demands coordination between molecular geneticists, breeders, and agronomists.

- *Genotype-Environment Interactions*: Agricultural traits are influenced not only by genetic factors but also by environmental conditions. Gene sequencing alone may not capture the full extent of genotype–environment interactions that affect trait expression and performance across different environments.

- *Identification of Functional Variants*: Gene sequencing can detect genetic variants (SNPs, indels, etc.) associated with traits of interest. However, distinguishing between functional and nonfunctional variants requires additional experimental validation and functional genomics studies. Not all detected variants directly contribute to phenotype variation.

- *Interpreting Noncoding Regions*: While gene sequencing is adept at identifying coding regions that produce proteins, noncoding regions of the genome still present challenges in terms of interpretation. Understanding the regulatory functions and interactions of noncoding DNA remains a complex and ongoing area of research.

- *Limited Functional Understanding*: Knowing the sequence of genes does not automatically provide insights into their functions. Functional genomics approaches are required to elucidate the roles of specific genes, which often involves additional experimental validations and functional assays.

- *Ethical and Regulatory Concerns*: The use of genetic information in agriculture and molecular breeding raises ethical issues related to genetic modification, intellectual property rights (patenting of genetic sequences), and regulatory approval processes and potential environmental impacts. These concerns can influence public perception and regulatory frameworks surrounding genetically modified organisms (GMOs) and gene-edited crops. Issues related to genetic discrimination, especially in healthcare and employment, are also significant considerations. Strict regulations governing the release and commercialization of genetically modified crops may delay their introduction to market.

- *Long-Term Environmental and Socio-Economic Impacts*: The introduction of genetically modified or gene-edited crops based on gene sequencing data may have long-term environmental and socioeconomic impacts. Assessing and mitigating potential risks to biodiversity, ecosystem services, and farmers' livelihoods is essential in adopting these technologies responsibly.

These demerits highlight the ongoing complexities and challenges associated with gene sequencing, necessitating continued technological advancements, interdisciplinary collaborations, and ethical considerations to fully harness its potential across different fields of science.

8. Conclusions and future outlook

In the last 20 years, plant genome sequencing has expanded quickly, increasing the amount and caliber of publicly accessible genetic information. The advances in human genomics technology have led to an enormous increase in our understanding of the human genome and its relation to disease [37]. Plant genomics, on the other hand, is a rapidly growing field essential for ensuring global food security. The past two decades have witnessed significant advancements in plant genome sequencing, genetic mapping, and the acquisition of biological data across multiple levels of biological organization. With the advent of new technologies and approaches, crop genome sequencing will continue to drive scientific progress in the years to come. Sequencing all plant species is a long-term goal that may become key to effectively supporting life on Earth through the improved management of plants in wild populations and their selection and genetic enhancement for use in agriculture and food production. Plant genome sequencing can also support adaptations to climate change by enabling the development of plant varieties and capturing a more comprehensive range of diversity from wild crop relatives and other plants that can help relocate agriculture and production in vertical farming. However, challenges remain, such as obtaining suitable DNA samples from plants, interpreting the large amounts of data generated, and identifying disease-causing environmental factors from genomic sequencing. Nevertheless, the abundance of genomic data offers a unique opportunity to better understand the biology and evolution of land plants, which have historically been dominated by affluent nations in the Global North and China. This knowledge can support conservation efforts, improve agricultural sustainability, and promote food and nutritional security. It is important to promote collaborative efforts and equitable distribution of resources to ensure that researchers from all nations have access to these valuable genomic resources.

Author details

Rhitisha Sood* and Vivek Singh
Department of Genetics and Plant Breeding, CSKHPKV, Palampur, India

*Address all correspondence to: rhitishasood5529@gmail.com

References

[1] Gupta AK, Gupta UD. Chapter 19 - animal biotechnology. Models in Discovery and Translation In: Next Generation Sequencing and its Applications. Academic Press, Elsevier; 2014. pp. 345-367

[2] Mardis E. Next-generation DNA sequencing platforms. Annual Review of Analytical Chemistry. 2013;**6**:287-303

[3] Petersen BS, Fredrich B, Hoeppner MP, Ellinghaus D, Franke A. Opportunities and challenges of whole-genome and - exome sequencing. BMC Genetics. 2017;**18**:14

[4] Zhou X, Ren L, Meng Q, Li Y, Yu Y, Yu J. The next-generation sequencing technology and application. Protein and Cell. 2010;**1**(6):520-536. DOI: 10.1007/s13238-010-0065-3/

[5] Aryal S. Next-Generation Sequencing (NGS) - Definition, Types. Microbe notes; 2022. Available from: https://microbenotes.com/next-generation-sequencing-ngs/

[6] Schadt EE, Turner S, Kasarskis A. A window into third-generation sequencing. Human Molecular Genetics. 2010;**19**:227-240

[7] Greenleaf WJ, Block SM. Single-molecule, motion-based DNA sequencing using RNA polymerase. Science. 2006;**313**:801

[8] Shendure J, Ji H. Next-generation DNA sequencing. Nature Biotech. 2008; **26**(10):1135-1144

[9] Rhoads A, Kin Fai A. PacBio sequencing and its applications. Genomics, Proteomics & Bioinformatics. 2015;**13**:278-289

[10] Deamer D, Akeson M, Branton D. Three decades of nanopore sequencing. Nature Biotechnology. 2016;**34**:518-524

[11] Goff SA, Ricke D, Lan TH, Presting G, Wang R, Dunn M, et al. A draft sequence of the rice genome (*Oryza sativa* L. ssp. japonica). Science. 2002; **296**(5565):92-100

[12] Brenchley R, Spannagl M, Pfeifer M, Barker GL, D'Amore R, Allen AM, et al. Analysis of the bread wheat genome using whole-genome shotgun sequencing. Nature. 2012;**491**(7426):705-710

[13] Rusch DB, Halpern AL, Sutton G, Heidelberg KB, Williamson S, Yooseph S, et al. The sorcerer II Global ocean sampling expedition: Northwest Atlantic through eastern tropical Pacific. PLoS Biology. 2007;**5**(3):e77

[14] Human Microbiome Project Consortium. Structure, function and diversity of the healthy human microbiome. Nature. 2012;**486**(7402): 207-214

[15] International Wheat Genome Sequencing Consortium (IWGSC). A chromosome-based draft sequence of the hexaploid bread wheat (*Triticum aestivum*) genome. Science. 2014;**345** (6194):1251788

[16] Tomato Genome Consortium. The tomato genome sequence provides insights into fleshy fruit evolution. Nature. 2012;**485**(7400):635-641

[17] Zhang Q, Li W, Li X, Liu G, Xia C, Xia Y, et al. The genetic architecture of wild rice adaptation and domestication revealed by whole-genome resequencing. Annual Review of Plant Biology. 2016;**67**:93-104

[18] Fracasso A, Trindade LM, Amaducci S. Drought tolerance strategies highlighted by two Sorghum bicolor races in a dry-down experiment. Journal of Plant Physiology. 2016;**190**:1-14

[19] Li C, Unver T, Zhang B. A high-efficiency CRISPR/Cas9 system for targeted mutagenesis in cotton (*Gossypium hirsutum* L.). Scientific Reports. 2017;**7**:43902

[20] Shi J, Habben JE, Archibald RL, Drummond BJ, Chamberlin MA, Williams RW, et al. Overexpression of ARGOS8 modifies ethylene responses, improves drought tolerance, and increases yield in maize. Plant Cell. 2017; **29**(10):2269-2281

[21] Sierro N, Battey JN, Ouadi S, Bovet L, Goepfert S, Bakaher N, et al. The tobacco genome sequence and its comparison with those of tomato and potato. Nature Communications. 2014;**5**:3833

[22] Qiao L, Zhang W, Qi Z, Zhou L, Wang X. Genome-wide identification and expression analysis of wheat bHLH transcription factor genes under various abiotic stresses. Environmental and Experimental Botany. 2019;**166**:103807

[23] Shi J, Gao H, Wang H, Lafitte HR, Archibald RL, Yang M, et al. ARGOS8 variants generated by CRISPR-Cas9 improve maize grain yield under field drought stress conditions. Plant Biotechnology Journal. 2019;**17**(8): 1470-1483

[24] Zheng GX, Terry JM, Belgrader P, Ryvkin P, Bent ZW, Wilson R, et al. Massively parallel digital transcriptional profiling of single cells. Nature Communications. 2017;**8**:14049

[25] Fang C, Ma Y, Wu S, Liu Z, Wang Z, Yang R, et al. Genome-wide association studies dissect the genetic networks underlying agronomical traits in soybean. Genome Biology. 2017;**18**(1):161

[26] Hamilton JP, Hansey CN, Whitty BR, Stoffel K, Massa AN, Van Deynze A, et al. Single nucleotide polymorphism discovery in elite North American potato germplasm. BMC Genomics. 2011;**12**:302

[27] Yu Q, Li B, Li X, Li Y. CRISPR/Cas9-induced targeted mutagenesis and gene replacement to investigate the function of water use efficiency-related genes in tomato. Horticulture Research. 2020;**7**:58

[28] Bamshad MJ, Ng SB, Bigham AW, Tabor HK, Emond MJ, Nickerson DA, et al. Exome sequencing as a tool for Mendelian disease gene discovery. Nature Reviews Genetics. 2011;**12**(11): 745-755

[29] Thirumalaikumar VP, Devkar V, Mehterov N, Ali S, Ozgur R, Turkan I, et al. RNA-seq and small RNA-seq analysis of heat tolerance in barley. BMC Genomics. 2021;**22**:394

[30] Villani AC, Satija R, Reynolds G, Sarkizova S, Shekhar K, Fletcher J, et al. Single-cell RNA-seq reveals new types of human blood dendritic cells, monocytes, and progenitors. Science. 2017;**356** (6335):eaah4573

[31] Bo K, Li Y, Zhang Z, Jia Z, Wang X, Zhang Z. The whole-genome sequencing and identification of candidate genes related to cold tolerance in cucumber (*Cucumis sativus* L.). Frontiers in Plant Science. 2020;**11**:1524

[32] Good AG, Beatty PH. Engineering nitrogen use efficiency for cereal crops: An integrated approach. Plant Physiology. 2022;**178**(2):5-13

[33] Silva TN, Santos JF, Borges AF, Carlini-Garcia LA, Melo AT. Genomic

regions associated with nutritional quality traits in tropical maize. Frontiers in Plant Science. 2023;**14**:1002345

[34] Tiwari JK, Devi S, Ali N, Sharma S. Whole-genome sequencing of potato identifies genomic regions associated with late blight resistance. Scientific Reports. 2023;**13**:23456

[35] Bolger AM, Poorter H, Dumschott K, et al. Computational aspects underlying genome to phenome analysis in plants. The Plant Journal. 2019;**97**:182-198

[36] Henry RJ. Progress in plant genome sequencing. Applied Biosciences. 2022;**1** (2):113-128. DOI: 10.3390/applbiosci1020008

[37] Marks RA, Hotaling S, Frandsen PB, et al. Representation and participation across 20 years of plant genome sequencing. Nature Plants. 2021;7:1571-1578. DOI: 10.1038/s41477-021-01031-8

Chapter 4

DNA Sequencing Technology Reveals Disparity in Mutagenesis Due to Fidelity Differences between Two Daughter DNAs in Evolution

Mitsuru Furusawa, Ichiro Fujihara and Motohiro Akashi

Abstract

Disparity mutagenesis, which focuses on the molecular basis of genetic information replication, is central to the disparity evolutionary theory. However, previous evolutionary theories have not fully addressed the molecular basis of replication. As a result, evolutionary simulations often incorrectly "assumed" equal mutation rates for both daughter strands derived from a parent strand, referred to as "parity mutagenesis" in contrast to "disparity mutagenesis." Multiple simulations have demonstrated that disparity mutagenesis has numerous unanticipated evolutionary benefits compared to parity mutagenesis. Molecular biological experiments have confirmed the imbalance in mutation rates among daughter strands, strengthening the disparity evolutionary theory. This review summarizes the existing studies on the disparity evolutionary theory and explores its future prospects. Furthermore, this report provides a comprehensive overview of the evolution of DNA sequencing technologies that facilitate the identification of disparity mutagenesis.

Keywords: asymmetry, DNA replication, evolution, disparity mutator, acceleration of evolution, lagging strand, punctuated equilibrium, genetic algorithm

1. Introduction

Darwin's theory of evolution, proposed in 1859, has remained alive till now. Such a long-lived theory is rare in natural science. The modern understanding of Darwin's theory can be roughly summarized using the following scheme. Germline mutations → inheritance (ontogeny) → phenotypic changes → natural selection → evolution. Among these, the primary factors for evolution are mutation and selection, the events that living organisms cannot positively control. Therefore, as observed from the viewpoint of a living organism, evolution is a passive event. Here, naive questions arise. (1) Is it really true that living organisms have no mechanism by which evolution is positively propelled? (2) Without this intrinsic propelling mechanism, is it possible to evolve from unicellular organisms to humans within just 4 billion years?

In the 1940s, Darwin's evolutionary theory and Mendelian genetics were integrated. Then, Neo-Darwinism was born, which has currently become a mainstream of biology. Population genetics began in the 1930s and has become a mainstream of evolutionary biology by embracing Neo-Darwinism.

The rough process of evolution proposed by population genetics may be expressed using the following scheme. Random mutations → genetic drift → natural selection → evolution. Other evolutionary factors may also exist, such as sexual isolation, sexual selection, gene flow, and genetic recombination. However, these additional factors are not crucial for the discussions presented here.

Our hypotheses on population genetics are indicated below. (1) As population genetics is founded on Darwinian evolution, it may be taking over the same problems concerning the intrinsic mechanism for propelling evolution, as mentioned above. (2) Population genetics is an academic field that studies dynamic changes in the population. Nevertheless, less attention has been paid to the molecular mechanism of DNA replication. DNA replication is fundamental in controlling cell divisions and forming a population. Above all else, the mutations introduced during DNA replication are one of the leading causes of evolution. (3) In population genetics, most studies have been conducted by hypothesizing that mutations are stochastically evenly distributed into two daughter DNAs. This presupposition may lead to the concept that mutations accompanying DNA replication are stochastically evenly distributed in chromosomes.

In classical population genetics, mutation rates remain low and are considered constant. The reason why the mutation rate was considered constant was due to the fact that the variation of the nucleotide sequence is constant, as reported by Kimura et al. in their neutral theory [1–3]. The idea that mutation rates are kept low also stems from the fact that the mutation rate of organisms is maintained at a low level of $10^{-7} \sim 10^{-10}$ (i.e., [4, 5]). Therefore, it is believed that the mutation rate remains low because of the trade-off between mutations that are beneficial to the host and those that are detrimental to the host as the mutation rate increases. Therefore, studies on the evolution of mutation rates based on classical population genetics have focused on neutral mutations, and the trade-off effects of fixed mutation rates on evolution have been examined [6]. In other words, the model of the relationship between mutation and mutation rate presented by classical population genetics is inductively derived from observational facts and has not been able to link DNA replication mechanisms and other factors within the cell.

In contrast, the disparity mutation model is an evolutionary model deductively derived from assumptions of the DNA replication mechanism. It was originally proposed as a method to solve the problem of mutation thresholds in Eigen's evolutionary study of RNA genomes ([7], Eigen is the Editor). As discussed below, many molecular biological experiments have reported that mutation rates vary between leading and lagging strands, depending on the DNA replication mechanism (see Section 4.5). In short, the theory of disparity evolution is an unusual model in that it is an evolutionary model that has advanced with the development of molecular biology, rather than as a model of classical population genetics.

If a causal relationship exists between DNA replication and the manner of entering mutations, we must reconsider the molecular mechanism of DNA replication. The structure of DNA replicating machinery was asymmetric—that was the start. Then, we assumed that this symmetrical breaking may cause disparity mutagenesis between two daughter DNAs. This review demonstrated that the fidelity difference between two daughter DNAs after replication increases or erases the so-called error threshold. Consequently, genetic diversity is increased far beyond our expectations, and evolution is accelerated. As it is generally believed, mutations and selection are the central players in promoting evolution. In addition, we propose the fidelity difference between two daughter DNAs as the third player for positively promoting evolution. This review provides a comprehensive summary of the 30-year-long research on disparity mutation effects. It examines the evidence from molecular biology experiments that reveal fidelity differences between the two daughter strands, supporting the existence of such disparities. Moreover, it establishes a solid groundwork for future and robust studies on disparity mutations within the field of evolutionary biology.

2. Reconsidering the DNA replication mechanism

We focused on how to replicate and transfer genetic information. DNA is a linear molecule comprising two nucleotide chains and has an anti-parallel structure regarding the direction of nucleotide-chain elongation. Each corresponding nucleotide is in a complementary relationship (A:T and G:C), and pairing nucleotides are connected with hydrogen bonds. The replication follows the semiconservative replication rule [8].

The double-stranded structure is unfolded via a hydrogen-bond cleavage at the beginning of replication. The resulting two parental single strands are then used as a template. DNA polymerase connects complementary nucleotides in the $5' \rightarrow 3'$ direction [8]. The polymerase cannot connect it in the opposite direction. This one-sided property of DNA polymerase raises essential issues, as discussed later.

Mutations are introduced when DNA replicates. The cause of the mutation is the accidental capturing of a nucleotide tautomer into the nascent DNA chain. Therefore, most mutations introduced are point mutations or single nucleotide replacements, resulting in a single nucleotide polymorphism formation among individuals. Thus, the point mutations introduced are one of the primary causes of evolution. As the existence

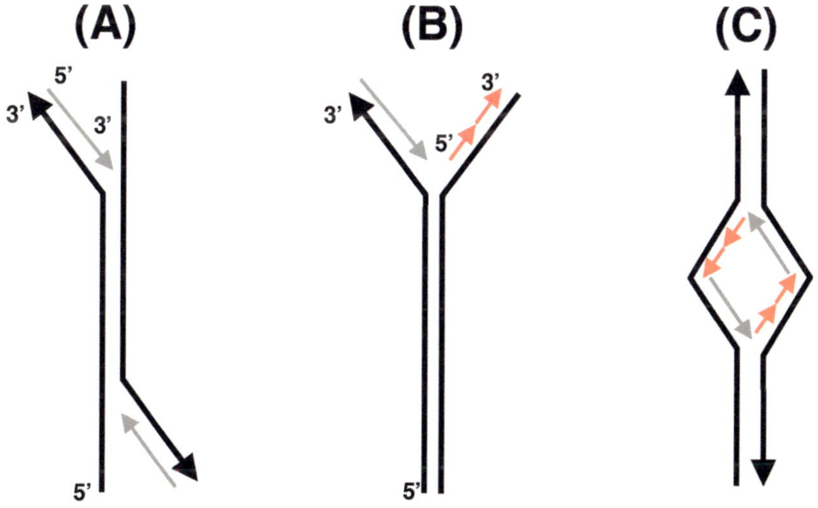

Figure 1.
Three plausible types of DNA replication. (A) Replication starts by using the continuous strand from both terminals of linear DNA. (B) From one terminal, replication starts by using continuous and discontinuous strands. (C) From the origin in the middle of a DNA molecule, replication starts by using continuous and discontinuous strands. Therefore, an eye-like structure is formed. A thin gray arrow indicates a continuous strand (leading strand). Discontinuous thin red arrows indicate a lagging strand.

of nucleotide tautomers is purely a matter of quantum chemistry, no organism can avoid the mutagenic effect of nucleotide tautomers. Therefore, no guarantee exists that accurate base pairs (A:T and G:C) are formed when DNA replicates. As long as a cell divides, it will change or evolve.

Judging only from this knowledge, a natural consequence would be that the mutation frequency is equally delivered into both daughter DNAs. As shown in **Figure 1A**, we imagined that linear DNA is replicated from both terminal sides simultaneously. As both strands may be synthesized similarly, there may be no difference in the average mutation rate between the two daughter DNAs. The occurrence of mutations during DNA replication is a random event in population genetics. At least, up until 1968, the situation was regarded as such.

3. Discovery of Okazaki fragments

The actual DNA replication mechanism in living organisms is not that simple. The length of a single chromosomal DNA strand necessitates the requirement of a significant amount of time for completion of replication when it is initiated from both ends (**Figure 1A**). Therefore, multiple replication origins (*ori*s) exist on a chromosome. Replication starts from each ori in both directions simultaneously. As a result, the DNA double-helix-chain becomes loose, and an eye-like structure is formed at the ori site (**Figure 1C**; for a detailed explanation, see the legend and text).

On the one hand, when DNA replicates, a new strand is synthesized as a continuous strand where the direction of replication coincides with that of the elongation of a new strand (leading strand synthesis). On the other hand, when the direction of replication is opposite to that of the elongation of a new strand, the new strand is discontinuously synthesized using fragments (lagging strand synthesis) (**Figure 1B**) [9]. These fragments were later named "Okazaki fragments." The DNA region synthesized by the lagging strand was initially regarded as having no biological activity. As it turns out, almost all genomic DNA is replicated by the leading/lagging system. Why the biological meaning of the Okazaki fragment did not become a central research subject is unknown, as more than 50 years have passed since Okazaki's discovery [10].

Certain crucial implications may emerge when any mechanism that requires a major cost has been placed in a long evolutionary process. In 1988, the present author (MF) thought that the fidelity of the lagging strand might be substantially lower than that of the leading strand. This lower fidelity is because the molecular mechanism for synthesizing the lagging strand could be much more complicated [11]. For example, when we compare two machines with the same purpose, a machine with more complicated systems might be error-prone [11]. Nowadays, this idea is not wrong [12–14].

Two reasons exist for our first paper on disparity mutagenesis to get published. First, the error-proneness of the lagging strand had not yet been discovered. Second, as this unbalanced mutagenesis substantially increases the error threshold, as noticed later, the refusal reason was "Something is wrong. The conclusion is against the fundamental law of thermodynamics."

4. Biological meaning of unbalanced (disparity) mutagenesis

When lagging-strand-biased mutagenesis occurs, rewriting a mutagenic landscape becomes necessary [12]. **Figure 2A** shows a typical mutagenic landscape in disparity mutagenesis. This landscape includes a pure DNA population comprising 100 individual DNAs. The individual linear DNA has one ori at its terminal and 100 nucleotides, corresponding to a replicore in an actual organism. This DNA population synchronously replicates once, and 200 daughter DNAs are produced. The mutation rate of the leading strand is one mutation per genome per replication (or 1%), and that of the lagging strand is 10%. Thus, the total mutation rate is 11%.

Figure 2B indicates the stochastic distribution of mutations derived from the leading strands of the parental population, including 100 DNAs. The mutations are shown according to the binominal distribution. Over 35% of the population has zero mutations, irrespective of the total mutation rate being 11% (**Figure 2B**). The distribution shows a bilateral symmetry in the binominal distribution for the lagging strand. The peak of the distribution is 10 mutations per DNA (**Figure 2C**). The mutagenic distribution of the total population is shown in **Figure 2A**, which can be obtained by combining the graphs of **Figure 2B** and **C**. The resulting graph consists of two peaks, the pattern of which is different from a binominal distribution (**Figure 2A**) [15].

Based on the abovementioned questions on population genetics, one of the present authors, MF, proposed the disparity theory of evolution [11, 12]: the fidelity difference between the leading and lagging strands causes an increase in the error threshold and the acceptable number of mutations in the population. Therefore, enlarged genetic divergence and robustness against environmental changes are expected. As a result, the promotion of evolution is realized.

Conservative individuals with few or zero mutations occupy about 40% of the population. This disparity system may have high robustness against mutations. If the

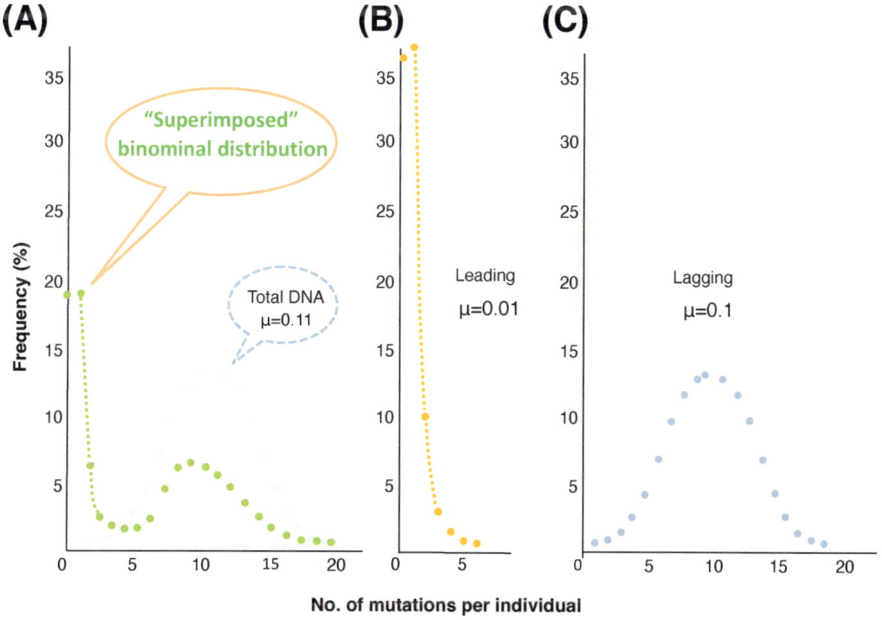

Figure 2.
Lagging-strand-biased mutagenesis causes a unique mutagenic landscape. The pure population of 100 DNAs replicates once. (B) Mutation distribution in the DNA population synthesized by the high-fidelity leading strand, mutation rate 1% (orange). (C) Synthesized by the low fidelity of lagging strand, mutation rate 10% (blue). (A) The mutagenic landscape of the total 200 daughter DNAs (green). Two peaks are observed: the arithmetic sum of (B) and (C) (This calculation was performed by M. Teraoka and T. Terasaki. Adapted from ***Figure 6****, Genetica, vol. 102/103, page number 339, 1998, Springer Nature.* ***Figure 6*** *of reference [15] with modifications.).*

environment is stable, the population may continue to survive by individuals with few mutations. We consider the situation where the total mutation rate is 11% in conventional parity mutagenesis. No wild type individuals may exist in the population. Thus, the robustness would be very low in the conventional parity mutagenesis model. In the disparity model, when the mutation rate of the leading strand tends to be zero with no limit, half of the population will be occupied by individuals with zero mutations.

Similarly, the disparity mutagenesis model of DNA explains the characteristics of living organisms, which encloses two conflicting potentials, that is, "things that do not change" (inheritance) and "things that change" (evolution). Although the evolutionary advantage of disparity mutagenesis is apparent, it becomes more evident when the pedigree of DNA is considered.

Figure 3 compares the parity mutagenesis model with the disparity one regarding pedigrees. For simplification, we focused on the replicore level because the replicore is the smallest replication unit of a chromosomal DNA. A chromosome comprises several replicores that are tandemly connected. The DNA shown in **Figure 3** corresponds to a single replicore (or replicon) in the actual chromosome. The characteristics of the disparity mutagenesis model are explained in the subsections below.

4.1 Increase in error threshold

The basic mutagenesis pattern would be unchanged when the mutation rate of the lagging strand is exclusively much higher than that of the leading strand (**Figure 3B**). Thus, theoretically, no error threshold occurs in this typical deterministic model because the original genotype with zero mutations can be guaranteed infinitely. In contrast, the conventional parity mutagenesis model only indicates the error-threshold situation (**Figure 3A**). The number of mutations per individual DNA continues

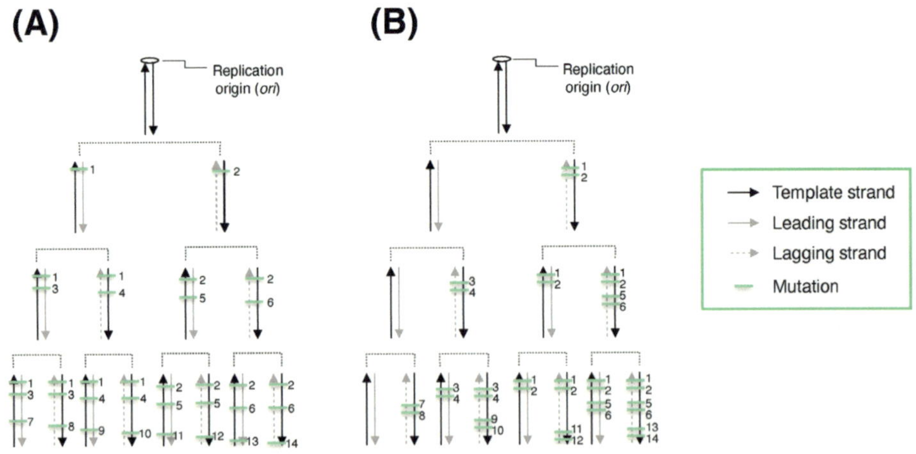

Figure 3.
Deterministic illustrations of the parity and disparity models for the mutagenesis accompanying DNA replications. (A) In the parity model, one mutation per replication is introduced in both daughter strands. (B) The disparity model introduces two mutations per replication exclusively in the lagging strand. A thin arrow indicates a leading strand, and a thin dotted arrow indicates a lagging strand. A short bar crossing strands is a mismatched nucleotide. Each number beside a bar is a base substitution at a different site. The ori indicates the replication origin (Adapted from ***Figure 3****, Genetica, vol. 102/103, page number 335, 1988, Springer Nature.* ***Figure 3*** *of reference [15] with minor modifications).*

to increase in proportion to the increase in generation times. As mutations are evenly distributed in every individual, this population will eventually become extinct by deleterious mutations. The disparity mutagenesis model predicts the rescue of living organisms from catastrophic errors. For review, see [14].

At the end of the twentieth century, a big controversy was triggered concerning this evidence; how can we survive? Several ideas were presented, such as the effect of population size, truncation selection, relative selection, and stabilizing selection. The present author belatedly participated in this controversy and argued the contribution of the disparity mutagenesis to this contradictory phenomenon [14].

4.2 Diversity enlargement with a principal guarantee

Figure 3B shows an extreme case of the disparity mutagenesis model. As the mutation rate of the leading strand is zero, the genetic information of the parental DNA is certainly inherited by one of the daughter DNAs. Then, the enlargement of diversity with the guarantee of "principal" (the parental genetic information) is attained. The term "principal" comes from economics. In other words, this mutagenic system guarantees that any genetic information, once it appears, will be inherited in perpetuity. For instance, DNA having "1" and "2" mutations, which first appeared in the second generation, emerged in the third and fourth generations (**Figure 3B**). The error-less leading strand might transfer this genotype to the descendant infinitely.

We showed that an appropriate recombination rate increased the fitness scores (FS) in the disparity system through a primitive DNA-type genetic algorithm (GA). The most effective recombination rate was approximately two per chromosome per replication [7]. This rate was the average value of recombination seen in natural organisms; excess re-combination decreased FS.

Thus, the evolutionary advantage of the disparity model at the replicore level would be apparent. An actual chromosome in eukaryotes has multiple oris. The effect of the multiple oris on the mutagenic landscape is complex. For example, DNA shown in **Figure 3B** is equivalent to a replicore. One of the two daughter DNAs is burdened with the entire length of the leading strand, meaning that this DNA has the same genotype as that of the parental DNA.

From a different viewpoint, the disparity mutagenesis model can protect essential genes from deleterious mutations accompanying DNA replication. One essential gene inevitably replicates using the leading or lagging strand, wherever it may exist on the chromosome. The error-less leading strand protects the gene from mutations independently of mutation rates (**Figure 3B**). This protecting ability is not affected by the size and position of the replicore nor by the number of oris. If multiple essential genes exist, recombination may increase the probability of a chromosome with the complete set of essential genes. In contrast, the conventional parity mutagenesis model cannot avoid the risk of deleterious mutations. This is because mutations accompanying DNA replication are randomly introduced throughout the genome, independently of the DNA strands (**Figure 3A**).

4.3 Theoretical grounding of the effect of disparity mutagenesis

The following studies are virtual experiments. Eigen et al. showed the existence of a critical error threshold using his "quasi-species" of DNA or RNA. The genome comprised of a binary base sequence of 50 [16]. The population size was large but finite. DNA polymerases with various fidelities were provided and separately added into a virtual reactor containing parental DNA and nucleotides. The number of mutations introduced in each DNA molecule was calculated when the replication reaction reached a stable state. When the error-less wild type enzyme was used, the reactor was entirely occupied by the wild-type master DNA (I0), which had the best fitness score among the possible nucleotide combinations. The mutant fitness (growth rate) was determined as 1/10 of that of the muster genome, irrespective of the number of mutations inserted. By decreasing the fidelity of the polymerase, the number of DNA molecules carrying more mutations increased, and DNA having 1, 2, 3 ... mutations occupied most of the population. If the polymerase mutation rate exceeded the critical value (approximately 2.3 mutations/DNA/replication), the population immediately fell into a chaotic sea (**Figure 4A**). The results mentioned here were for the parity mutagenesis model (c = 0).

Different results were obtained when a mixture of error-less and -prone polymerases were used. By increasing the error-free polymerase ratio, the position of the error threshold moved to the right (c = 0.09) and finally disappeared (c = 0.1). When a mixture of 30% error-free and 70% error-prone polymerases were used, the extinction of the population was avoided, even if the average mutation rates increased. Notably, the wild type with zero mutations (I0) existed forever as far as examined (**Figure 4D**). We calculated the maximum mutation number per replication capable

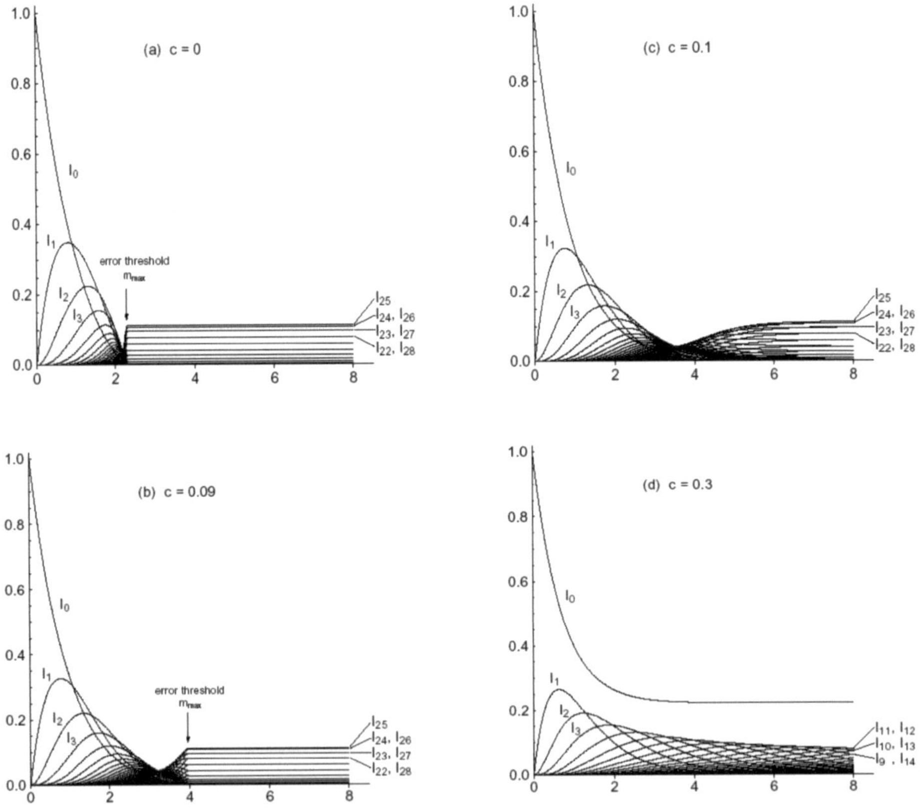

Figure 4.
Mutant distribution in "quasi-species" is a function of the mean error rate per genome. The genome comprises a binary base sequence of 50. (In); the sum of the relative stationary concentration of the sequence with n-mutations. (Io); all zero-mutations or wild type are plotted. c = 0 is the parity mutagenesis model. c = 0.3; 30% of wild type error-free polymerase. The arrow indicates the error threshold. A single polymerase molecule contributes to synthesizing the entire genome length (Adapted from ***Figure 1*** *of the reference [17]).*

of being introduced into a single *Escherichia coli* cell. The error threshold of the parity system was 2.3 mutations/cell/replication, and that of the disparity system was 31. These effects of the mixture of DNA polymerases with different fidelities can correspond to the effects of lagging-strand-biased mutagenesis in DNA [17].

4.4 Possibility of experimental acceleration of evolution

When we try to increase mutation rates, we consider treatment with mutagenic chemicals or using irradiating electromagnetic waves, such as UV. However, these methods cannot accelerate evolution because these physicochemical treatments cause indiscriminate mutations that may cause severe deleterious effects on DNA, RNA, and proteins.

The concept of disparity mutagenesis would provide us with a different methodology for realizing evolutionary acceleration. We may experimentally accelerate evolution by using a specific mutant, which keeps intact low mutation rates in the leading strand and has an appropriately high level of mutation rates in the lagging strand. This mutant having unbalanced mutagenesis between two strands is called the "disparity mutator" after this [15]. The evidence of the acceleration of evolution will be presented through a genetic algorithm and laboratory organisms.

4.4.1 Evidence of disparity mutations from molecular biology experiments

Recent molecular biology experiments have supported the existence of disparity mutations between the daughter strands. Initial evidence of the mutation imbalance was obtained using plasmid-based methods, which compared the mutation rates of drug resistance genes in different coding directions relative to the origin of replication. A high mutation rate was confirmed in the lagging strand [18]. Further studies using the same method on bacterial genomes supported this hypothesis and attributed the high mutation rate to the collision and repair of DNA replication and transcription [19]. However, some studies have reported that mutations in the *E. coli* genome are more frequent in the leading strand than in the lagging strand [20], whereas others, using next-generation sequencing, found that mutations in the Polα-replicated region accounted for approximately 1.5% of mutations in the lagging strand [21]. Tracking the genomic mutational footprint of a Polδ mutant (L612M, a proofreading activity-deficient strain) that replicates eukaryotic lagging strands, it was reported that it is biased toward mutations that Polδ is prone to [22]. A study on replication enzyme mutants in human cancer cells reported that Polδ, which replicates the lagging strand, is more mutated than Polε, which replicates the leading strand, and that the mismatch repair (MMR) machinery is activated at a three-fold higher rate in the lagging than in the leading strand [23]. The MMR machinery reduces insertion and deletion mutations in the mutational bias caused by the different enzymes replicating the leading and lagging strands [24]. The ribonucleotide repair of the leading and lagging strands of *E. coli* is unequal, with RNase HII responsible for ribonucleotide excision repair (RER) of the leading strand, whereas other repair systems such as RNase HI and nucleotide excision repair being involved in lagging strand repair [25]. The authors particularly emphasized that RNase HI activity affects the repair of single ribonucleotides incorporated into the lagging strand by replicase pol III HE. Polymerase usage sequencing (Pu-seq) developed by Koyanagi et al. was adapted to HCT116 cells and revealed genome-wide usage of Polε and Polα, suggesting that transcription-induced uncoupling of these enzymes during

Polε and Polα lagging chain replication is associated with chromosomal instability [26]. The uncoupling of these replication enzymes induced by transcription when Polε and Polα replicate lagging strands is associated with chromosomal instability [26]. Structural and functional analysis of the interaction between human Polα and the template-primer mismatch showed that in the presence of dNTPs, Polα selected the correct template 10-fold more frequently than the template-primer mismatch [27]. Additionally, Polα elongated the T-C template-primer mismatch 249-fold less efficiently (compared to the correct template), and the template-primer mismatch affected both substrate recognition and DNA elongation by Polα. In yeast, the mutation rates of the leading and lagging strands are equal for the calibration activity and MMR by Polδ [28]. Although these results seem to be at odds with those in humans and *E. coli*, they suggest that the mutation repair rates of the leading and lagging strands by proofreading activity and MMR produce final inter-organism diversity in the mutation rates of the leading and lagging strands, given that the replication accuracies of the leading and lagging strands are originally different. In the same yeast study, mutational analysis in single cells has shown that the mutation rates vary from one cell to another, and that the inheritance of mutations follows the rules of semi-conservative replication and mitosis [29]. Protein modifications also affect replication accuracy, and Polδ acetylation leads to increased replication accuracy [30]. Genetic asymmetry is also considered to exist for the methylation state of chromatin [31]. There are other factors other than DNA replication errors that create an imbalance in mutational tendencies between daughter strands. For example, stalling replication forks, which causes mutations on the lagging strand, can trigger an increase in insertion mutations owing to transposon activation [32]. Transposons and endogenous viruses may also create an imbalance in mutational trends between daughter strands. The error-prone bypass of DNA lesions during lagging-strand replication is a common source of germline and cancer mutations [33]. Regardless, the results of these experiments suggest that the two daughter strands have different mutation rates and propensities, constituting evidence of disparity mutations. Although the inherited mutations depend on the repair rate and mutation repair mechanism tendencies such as MMR, many experimental results still support the concept of disparity mutagenesis, even after accounting for these factors.

4.4.2 History of DNA sequencing technology

Looking back, it is no exaggeration to say that the analysis of the disparity mutation has been aided by the development of DNA sequencing analysis technology. Therefore, we will look back at the development of DNA sequencing analysis techniques.

The first method to appear was the combination of dideoxy and capillary electrophoresis. This was followed by the development of the first generation of DNA sequencing methods, the Maxam–Gilbert and dideoxy methods, and then the ABI 370 automated DNA sequencer from Applied Biosystems in 1987, which was also used to determine phage, bacterial, and human genomes and is still in use today.

By 2007, second-generation DNA sequencing methods had emerged, such as Roche's Roche-454 GSFLX+, which uses pyrosequencing, and Solexa's Genome Analyzer, which uses sequencing by synthesis (SBS). SBS is a method that uses a single nucleotide sequence to identify base pairs in the genome. The Genome Analyzer was capable of outputting 1Gb of sequence data in a single run when it was launched in 2006. Subsequently, Solexa was acquired by Illumina, and a series of more powerful sequencers using SBS were launched, including the HiSeq 2000 (2010), MiSeq

(2011), and NovaSeq (2017); the NovaSeq X series (2022~) has a capacity of 16 Tb. Note that so-called next-generation sequencing (NGS) comes after this era.

The Ion PGM™ sequencer is a sequencer introduced in 2010. The principle is similar to pyrosequencing, but emulsion PCR is performed on a semiconductor chip, and the released hydrogen ions are measured and sequenced with an on-chip pH meter. Unlike SBS, no optics are used, making the instrument compact. Ion PGM was acquired by Life technologies in 2010, and Life technologies was acquired by Thermo Fisher Scientific in 2015, but today Ion Torrent NGS systems continue to offer automated nucleic acid extraction and library preparation.

The third-generation, single-molecule real-time sequencing, SMRT sequencing, was introduced by Pacific Biosciences in 2011. The long-read sequencing method was problematic when first introduced due to the high number of errors, but this was greatly improved with the introduction of the HiFi (HiFi: high fidelity) sequencing method in 2019 [34]. This method, also known as Circular Consensus Sequencing (CCS), cycles the DNA to be decoded and reads repetitive sequences, allowing the detection of various mutations such as substitutions, deletions, and insertions with over 90% accuracy, while minimizing the effects of GC bias and repetitive sequences.

The fourth generation are the nanopore sequencers (MinION, GridION, and PromethION) from Oxford Nanopore Technologies, which are also long-read sequencers that determine sequences on a molecule-by-molecule basis. The sequence to be decoded is determined by measuring the change in electrical potential that occurs when ssDNA passes through nanopore proteins embedded in the membrane. The membrane containing the nanopores is constantly under voltage and the change in current caused by the obstruction of ion flow as the DNA passes through is measured. Sequences can be decoded by 1D analysis, which reads only one strand, or by $1D^2$ analysis, which reads both the major and minor strands, with an error rate of a few per cent in the latter case. Today, the application of nanopore-based DNA sequencing technology makes it possible to comprehensively determine the amino acid sequence of a protein [35].

4.5 Accelerating evolution using the disparity mutator

To accelerate evolution using living things, we proposed a specific mutant, a disparity mutator, in which the lagging-strand-biased mutagenesis was achieved [14]. In *E. coli*, the leading and lagging strands are synthesized by the same enzyme, and another gene product, *dnaQ*, conducts proofreading. When the activity of *dnaQ* is removed, the low intrinsic fidelity of the lagging strand, which is due to its mechanical complexity, appears. As a result, lagging-strand-biased mutagenesis occurs [18].

In eukaryotes, both strands are fortunately synthesized by different enzymes. The lagging strand is exclusively synthesized by *pol δ*, which has a proofreading domain. When the critical amino acid(s) for the proofreading is replaced by alanine, we can obtain the mutant that performs the lagging-strand-biased mutagenesis. To make a rough estimate, average mutation rates were increased by 1–2 orders of magnitude of the expected mutation rate in *E. coli* [18] and 17–18 times higher than the expected mutation rate in mice [36], as this type of disparity mutator accumulates point mutations exclusively into the lagging strands. The critical point is that the excess point mutations introduced are due to the incorrect capture of the tautomer of nucleotides, as seen in wild type organisms. Accordingly, we might be able to accelerate evolution according to the manner occurring in nature. As *pol δ* is well preserved in eukaryotes, disparity mutators can be produced using the same genetic engineering methodology.

We could obtain a mutant with an intended trait by controlling environmental conditions. A desirable trait can be fixed by replacing the mutant *pol δ* proofread with the wild-type *pol δ*. However, keeping the three-dimensional structure of *pol δ* intact would be necessary as far as possible. Changes in the molecular structure of the polymerase by genetic manipulations might cause crucial damage to DNA replication.

We could prepare disparity mutators in *E. coli*, yeasts, malaria parasites, and mice in another laboratory (see [11]) for review). For instance, we could breed an *E. coli* mutator that produced large colonies in the presence of saturated antibiotics. The mutated strain with strong tolerance to the given antibiotics used as a selection pressure did not show considerable tolerance to other antibiotics. In other words, the disparity mutator exclusively adapted to the antibiotics used as a selection pressure [11].

A heterotrophic bacterium, *Sphingobium japonicum*, cannot grow without an organic carbon source. However, *S. japonicum* proliferated even under deplorable nutritional conditions. A disparity mutator of *S. japonicum* could grow when the *adhX* gene was mutated with hyper-production of alcohol dehydrogenase. Surprisingly, this mutant can grow without adding organic carbohydrates. This finding may mean that this mutant can fix CO_2 in the air using an unknown pathway without chlorophyll [37]. In this system, as a wild-type proofreading gene and the mutant proofreading gene coexisted in a bacterial body, we could imagine a similar situation shown in **Figure 4D** might be realized [17].

In a malarial parasite, a known drug-resistance gene was identified [38]. In budding yeasts, we could identify two new temperature-sensitive genes, including hot 1; the concurrence of these genes contributed to tolerance at 40°C [39].

The growth rate of the disparity mutator is equal to that of the wild type. This feature is advantageous for the experimental acceleration of evolution since the polymerase function could be kept intact. The growth delay must function as a factor against the accelerating evolution. Even more evolutionary acceleration may be possible when the combined usage of the *pol δ*-proofread with appropriate mismatch-repair-deficient gene(s) is considered. Most importantly, the lagging-strand-biased mutagenesis was observed even in wild-type organisms [18], and this unbalanced mutagenesis mainly was owing to the mechanical complexity in the lagging strand synthesis, as we predicted before [11, 14].

A study with contrasting results to ours regarding the fidelity of DNA strands was published. The authors showed, using *E. coli*, that the fidelity of the lagging strand was higher than that of the leading strand [18]. We also used *E. coli* and showed that the leading strand has higher fidelity [18]; this may be because we used plasmid, whereas they used genomic DNA. The different results may be because of the different experimental systems used. Therefore, elucidating the exact mutation rates with asymmetrical DNA replication in an intact living organism is challenging because mutations observed in a genome are the sum of the mutations by lesions and those by replication errors. Therefore, this problem remains to be resolved.

The characteristics of mutagenesis using disparity mutators are summarized below.

1. There is a natural limit to clarifying the relationship between mutations and phenotypes by the comparative study of living organisms. This limit is because the effect of a newly introduced mutation on the phenotype is influenced by the phenotype when the mutation is precisely introduced. The phenotypic difference observed between the two species is due to countless mutations introduced in the past. In contrast, the disparity mutator method enables us to see the direct effect of mutations on phenotypic changes.

2. Even when using the disparity mutator method, we cannot imitate the precise course of future evolution because no one knows when and where mutations will be introduced. However, the disparity mutator method might reveal the inherent evolutionary potentiality of the organism.

3. The disparity mutator can be helpful as a novel breeding method. When a disparity mutator shows a desired phenotype, you can fix it by replacing the mutator gene with the wild type one.

5. Extended interpretation of the disparity theory

The pedigree of single-stranded RNA (ssRNA) and DNA (ssDNA) also guarantees the principal as the disparity model of DNA. The original ancestral RNA is a single plus strand (F0) (**Figure 5**). ssRNA viruses encode genetic information in the positive or negative strand of the virus particle. When the RNA replicates, a complementary minus strand is synthesized once in the opposite direction, and a temporal double-stranded molecule like dsDNA is produced. The temporal RNA/RNA double-stranded molecule dissociates, and a novel plus strand (F1) is synthesized using the minus strand as the template. Here, one replication cycle is completed. As the intact original plus strand (F0) remains in the system, the principal genetic information is completely guaranteed. In addition, every RNA molecule represented in **Figure 5** might be theoretically kept intact without exceptions. The strong infectivity of RNA viruses may depend on this unique replication manner of a single-stranded RNA. As stated above, the effect of disparity mutagenesis is most typically realized in ssRNA and ssDNA. Thus, the robustness of the RNA pedigree is similar to that of the DNA pedigree in **Figure 3B**, where the mutation rate in the leading strand is zero.

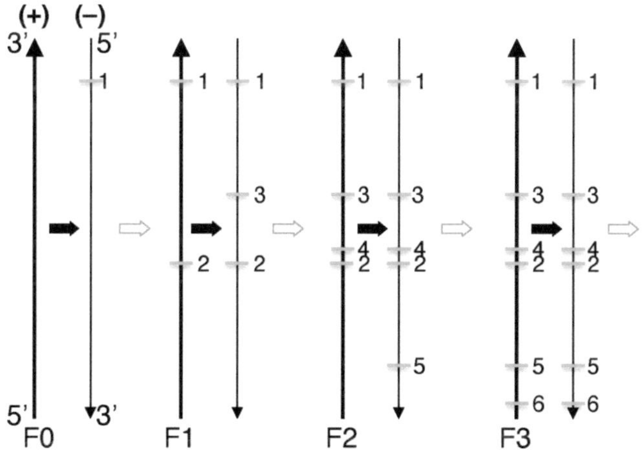

Figure 5.
Deterministic illustrations of accumulating replication errors in the pedigree of a single-stranded RNA. The original plus-strand RNA (F0, +) is copied as a minus strand (−) by RNA-dependent RNA polymerase. The minus-strand template RNA is copied as a daughter plus-strand RNA in the F1 generation. This cycle is repeated. A thin horizontal short bar with a number is a point mutation, and the different number indicates a different mutation. Every mutation, once introduced, is inherited forever in this population. For a detailed explanation, see Section 5.

DNA has an innate weak point that tends to be overlooked. Unlike ssRNA and ssDNA, parental DNA has to simultaneously disappear as replication completes. Therefore, when mutations are evenly distributed into two daughter DNAs, as seen in **Figure 3A**, the principal of genetic information is not necessarily guaranteed. Therefore, regarding the robustness against environmental changes, ssRNA and ssDNA gain an advantage over dsDNA. This nature would be a major reason ssRNA and ssDNA viruses are hard to eliminate and can easily cause pandemics.

6. 2-dimensional genetic algorithm

We aimed to devise the genetic algorithm (GA) closer to a natural chromosome; therefore, we developed the 2-dimensional genetic algorithm (2D GA). The novel GA comprised of 50 functional genes, and 200 junk DNA were evenly distributed between functional genes. All genes interacted with each other. The disconnection of gene interaction caused by mutations led to death. Every functional gene had three innate fitness landscapes. The population comprised of 100 homogeneous individuals when the simulation started. The population increased by synchronous replication. When the population size exceeded approximately 2000, the excess individuals were removed from the population by truncation selection. The genetic diversity of a population (Shannon–Wiener H′) was also considered [40]. The results and conclusions obtained are described below.

6.1 Conditions for quick adaptation

We first present the results of the simulation (**Figure 6**). The speed of adaptation can be controlled by varying the width of the difference in fidelity between the leading and lagging strands. The optimal conditions for rapid evolution were as follows: In the disparity system, the mutation rate of the lagging strand (*Pa*) was between 2 and 4, and the mutation rate of the leading strand (*Pe*) was less than 0.1. If *Pe* is close to 0, the population will not go extinct even if *Pa* is close to 10, indicating that the population is stable with some *FS* and a small H′. By contrast, in the parity system, extinction occurs when $Pe \approx 1$. Also, the conditions under which stable species were generated were very narrow, only close to $Pe = Pa = 0.65$ (data not shown, see reference [40]).

6.2 Differential equation for evolution with disparity mutagenesis

From the results of simulation experiments using 2D GA, we obtained two differential equations for the evolution of parity and disparity systems under the following assumptions: Suppose that two cells arise from one cell. It is assumed that the following two conditions are more favorable for a generational change to leave a larger number of mutations to the next generation.

1. After cell division, many cells were survived.

2. After cell division, many mutations were present in surviving cells.

Therefore, we have defined the activity coefficient of evolution, $a(Pa, Pe)$ ($a(Pe)$ in the parity system), which is shown in the following equation.

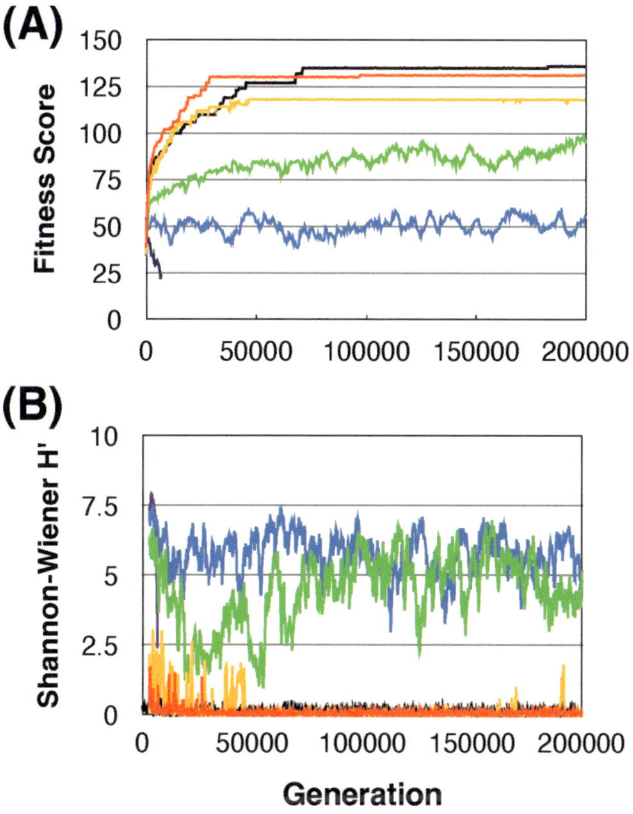

Figure 6.
Effect of the fidelity difference between the lagging and leading strands on evolution in the disparity model; Pe *dependency on evolution when* Pa *= 3. (A) The relationship between generation and fitness score. (B) The relationship between generation and genetic diversity (Shan-non-Wiener H′).* Pe *values: black line (*Pe *= 0.0), red (0.03), orange (0.09), green (0.12), blue (0.135), and dark blue (0.15). (Diverted **Figure** 5 from reference [40]).*

$$A_{disp}(Pa, Pe) = \left[(1-Pd)^{Pa} - Pe \times Pd\right] \times \left[Pa \times (1-Pd)^{Pa} + Pe \times Pe(1-Pd)\right] \quad (1)$$

$$A_{pari}(Pe) = (1-2Pe \times Pd) \times \left[2Pe \times (1-Pd)\right] \quad (2)$$

$$FS = F_{mx} \times a_{pari}(Pe) \times (1-\exp[-C \times \{1-2Pe \times Pd\} \times \{2 \times \{Pe \times (1-Pd)\}] \times G) \quad (3)$$

$$FS = F_{mx} \times a_{disp}(Pa,Pe) \times \left(1-\exp\left[-C \times \left\{(1-Pd)^{Pa} - Pe \times Pd\right\} \times \left\{Pa \times (1-Pd)^{Pa} + Pe \times (1-Pd)\right\}\right] \times G\right) \quad (4)$$

a (Pa, Pe): Normalized activity coefficient of evolution.
FS: Fitness score of the cell.
Fmx: The largest *FS* value.
G: Cell generation.
Pa: Mutation rate of the lagging strand (integer ≥ 2).
Pe: Mutation rate of the leading strand (real number < 1, $Pa = Pe$ in the parity system).
Pd: Probability of cell death due to a single mutation, 0.45.

In Eqs. (1) and (2), the first term is the average number of cells that survive the generation change, and the second term is the average number of mutations that enter the surviving cells. The differential equation for the generation changes of *FS* is formulated as the rate of generation change of FS, and the maximum possible values of *FS* are proportional to $a(Pa, Pe)$ (or $a(Pe)$). Eq. (3) is the generation dependence of the *FS* for the parity system, and Eq. (4) is for the disparity system.

The *Pd* and integration constant *C* used in Eqs. (3) and (4) were not fitted to *Pa*- and *Pe*-specific data but were applied to the whole system, calculated using values obtained from the relationship between *Pa* and *Pe* when the species became extinct. Both equations reproduce the results of the simulation experiments for both systems very well [41]. The specific simulation methods and derivation of Eqs. (1)–(4) are described in detail in the original paper Refs. [40, 41] and should be referred to.

These equations are not a function of time but of the number of generations (*G*). There are two variables: the mutation rate of the leading strand (*Pe*), and the mutation rate of the lagging strand (*Pa*). In Eq. (2), if $Pe \approx 0$, the population will not become extinct even if the value of *Pa* exceeds 10. In other words, when the error threshold disappears or becomes impossibly small, the population will not self-destruct owing to excessive genetic load. This conclusion can be understood intuitively from **Figure 3B**.

6.3 The relationship between the neutral and disparity evolutionary theory

Our study of the evolutionary process using the 2D GA model has provided crucial insights into the features of disparity mutations and their relationship to the neutral theory of molecular evolution. Our research has revealed two main findings [42]. First, gene-gene interactions are critical to evolution, restricting genetic diversity and preventing the excessive diversification of individual population members. Second, we identified three evolutionary modes: Stun, Evolution (Evo. mode), and Diversification (Div. mode). In Stun mode, the population size remains unchanged despite a lack of increased fitness, whereas in Evo. mode, fitness improves. In Div. mode, genetic diversity occurs while fitness remains relatively constant. Notably, Stun mode is specific to disparity mutations and helps to preserve gene-gene interactions by preventing excessive mutation rates, thereby maintaining the stability of the population's phenotype without advancing its adaptation level. Note that in the 2DGA evolutionary simulations, we attempted to simulate evolution with the same mutation rate and different gene cluster sizes using different random seeds, and in many cases, the same mutation rate resulted in slightly different increases in fitness (Brunner-Munzel and Kolmogorov-Smirnov tests, $p < 0.05$), but the values were similar for different gene cluster sizes [42].

The neutral theory of evolution states that the accumulation rate of synonymous substitution mutations in a gene is constant [43, 44]. The disparity mutation model appears to encompass the neutral theory, as shown in simulations using the 2D GA model [42]. In these simulations, the genetic distance of a population reaches a fitness plateau that is nearly constant and small, regardless of parity or disparity mutation conditions [42]. Mutations that occur when the adaptation level reaches a plateau do not substantially alter the adaptation level and may even reduce it, leading to a slow change in fitness. This situation is similar to that described in the neutral theory, referred to as the "diversification mode." In the neutral theory of molecular evolution, the mutation rate per time is constant, whereas, in our simulations, the mutation rate per generation is constant [41]. The constant mutation rate per time is considered valid when the number of cell divisions and proliferations per cell is constant among species for a specific gene being compared.

The disparity mutagenesis would contribute to proficiently promoting evolution. Genomic DNA may contain much concealed genetic information, such as the unknown concealed genetic information that is not expressed at present, pseudogenes, and recessive genes. This genetic information has the potential to change suddenly in promising genes by the addition of a few mutations. These incomplete reserve genes must be protected in the population for the expected evolution jump. The disparity mutagenesis model might protect them by the high fidelity of one of the daughter DNA (e.g., leading strand). However, the parity mutagenesis could not effectively protect them owing to the evenly introduced mutations. Extremely very low mutation rates could protect them; instead, we might give up the future quick evolution.

7. Discussion

Unlike other biological sciences, the theory of evolution is challenging to prove experimentally. Approximately 30 years ago, we developed a testable evolutionary theory in the laboratory. Returning to the starting point, we tried to factor in analyses for evolution: (1) Evolution occurs as a population. (2) The population is formed by replicating individuals that comprise the population. Replication mutations are one of the primary evolutionary factors. (3) The replication of individuals is controlled a priori by DNA. (4) The coupling of DNA replication and errors in DNA replication is the key to solving the evolution question.

In the past 30 years, no evolutionary theory that considered the molecular mechanism of DNA replication was presented. Genomic DNA is replicated by using the leading and lagging strands without exceptions. Most experiments showed higher mutation rates in the lagging strand. However, as stated in this review, the "disparity effect" can be expected when the fidelity difference between both daughter DNAs exists irrespective of the leading and lagging strands. In the disparity mutagenesis world, the effect of mutations is not considered as the simple sum of the mutation rates of both daughter DNAs. Counterintuitively, when the fidelity difference between two daughter DNAs exists and the mutation rate of one side of the DNA is low enough, the error threshold becomes higher or disappears. Therefore, in theory, the population can be permitted mutations without limitation. We can see this situation in the animal world, where many animals have far greater numbers than one mutation/genome/generation (for review, refer to [14]).

Evolutionary models based on population genetics are based on individual-based observations and have the weakness that models can only be developed and refined from currently observable differences. Indeed, the adaptation of this "random and normally distributed mutation," as described in population genetics, to the RNA replication and evolution model led Manfred Eigen to discover the problem of the mutation threshold, or "Eigen's paradox," and its inconsistency in evolution [45]. The disparity model solves this mutation threshold problem by assuming different mutation rates for the two daughter strands, which are mechanically inherent in the DNA replication mechanism, and the manner in which the model is constructed differs from the classical population genetics model. As noted above, recent validation by molecular biology experiments supports disparity mutation theory, making it an observable fact rather than a theory. The impact of disparity mutations on evolution, as with other evolutionary models, is currently difficult to test, because it takes an extremely long time. On the other hand, the strength of this model is that it is based on certainty at the molecular level, so it can detect and test minute and chaotic phenomena that would not be detected by evolutionary theories derived from population genetics.

Since the disparity theory was proposed, its certainty has been suggested through various attempts. The existence of disparity mutagenesis has been almost sufficiently ascertained by molecular biological approaches (see Section 4.5.1). However, in natural organisms in which a chromosome has several replicons and the existence of sexuality and crossing over, further studies remain to be conducted. The analyses using data on mutations may reinforce the theory of disparity mutagenesis.

Our original problem of the acceleration of evolution at the laboratory level can be accomplished using a "disparity mutator" [14]. As summarized in Section 4.5, disparity mutators of unicellular organisms were prepared by deleting the proofreading activity of DNA polymerase for the lagging strand. In disparity mutators, the fidelity of the lagging strand was exclusively decreased, and our initial goal, the acceleration of evolution, was attained. However, experiments with higher organisms remain to be conducted.

Darwin and Mendel could not realize disparity mutations because gene information did not exist then. However, things are different in population genetics. Population genetics started in the 1930s and currently comprehends Darwinian evolution, Mendelian inheritance, and molecular genetics. Population geneticists had not presented the disparity mutagenesis effect until we published our research in 1992. Most population geneticists have almost ignored our idea for more than 30 years. We would imagine that this is simply because of their oversight of the biological meaning of Okazaki fragments. Any active biological meaning of the Okazaki fragment had also not been found by molecular biologists, including the Okazaki group. Currently, paleogenetics enables the study of past DNA changes and its contribution to human evolution. In addition, we can predict the future image of a given dynamic system using short-term biological evolution experiments or computational biology. Based on the development of evolutionary models, conventional evolutional genetics was reached. However, these models could have been developed when required. Initially, our concept of disparity mutagenesis was exclusively applied to dsDNA replication. However, we can now use this concept for broader interpretation; this includes the principle of general biological replication processes. A limitation of the disparity mutagenesis theory, which is similar to other evolutionary theories, resides in our inability to determine the ultimate goal or trajectory of evolution, as the occurrence and nature of future mutations remain unknown. To advance research in disparity mutagenesis, our primary approach involves accumulating a substantial body of evidence through numerous experiments next-generation sequencing (NGS) and simulations.

In order to detect mutational bias in the leading and lagging strands using NGS technology (see Section 4.5.2), a number of different yeast genotypes were employed [21, 22]. The value of the methods employed in conjunction with NGS lies in their capacity to distinguish between mutations on the leading and lagging strands. With regard to disparity evolution, the next objective is to examine the impact of mutational bias on the leading and lagging strands on evolutionary processes in detail. The utilization of single molecule sequencing technologies with genome DNA labeled with nucleotide analogues will facilitate future analyses. In this context, experimental evolution is an appropriate approach for analysis.

Owing to the remarkable progress in computing machinery, we can construct a highly efficient evolutionary model and analyze the data. Future evolution genetics is believed to be: (1) a model-leading study and (2) a biological-data-driven study. This two-way strategy would reveal that disparity mutagenesis might be the main driving force of evolution.

8. Conclusion

The disparity mutation model was proposed in 1992 and has since been tested directly and indirectly in molecular biology experiments and computer simulations. This model is unique in that it is deductively derived from the complexity of replication of the leading and lagging strands of the DNA replication machinery and the error-prone nature of complex systems, unlike models based on the measurement of mutation rates used in population genetics. Therefore, it is more compatible with game theory and ecological models than population genetics. In recent years, molecular biology experiments have not been conducted to test the model, but to clarify the effects of pure DNA replication mechanisms on mutation, have solidified the unbalanced mutation model as an evolutionary system supported by a body of facts rather than a hypothetical model. However, as with general evolutionary models, one of the limitations of this study is that it is difficult to test with long-term observations. Our recent focus has been on the effects of unbalanced mutations on evolution in the presence of gene-gene interactions, which can avoid extinction due to the accumulation of deleterious mutations and produce diverse genotypes. Future work is needed to determine the differences in the results of implementing disparity mutation models in population genetics, ecology, and game theory problems.

Acknowledgements

We would like to thank Dr. F. Rueker for reading the manuscript and Dr. T. Yagi for his helpful suggestions. We would also like to thank Editage (www.editage.com) for English language editing and helpful comments.

Conflict of interest

The authors declare that the research was conducted in the absence of any commercial or financial relationships that could be construed as a potential conflict of interest.

Other declarations

This research received no external funding.

Author details

Mitsuru Furusawa[1], Ichiro Fujihara[2] and Motohiro Akashi[3*]

1 Chitose Laboratory Corp, Biotechnology Research Center, Kawasaki, Japan

2 College of General Education, Osaka Sangyo University, Osaka, Japan

3 Faculty of Science and Technology, Department of Science and Technology, Seikei University, Tokyo, Japan

*Address all correspondence to: motohiro-akashi@st.seikei.ac.jp

References

[1] Sueoka N. On the genetic basis of variation and heterogeneity of DNA base composition. Proceedings of the National Academy of Sciences of the United States of America. 1962;**48**(4):582-592. DOI: 10.1073/pnas.48.4.582

[2] Kimura M. Evolutionary rate at the molecular level. Nature. 1968;**217**(5129): 624-626. DOI: 10.1038/217624a0

[3] King JL, Jukes TH. Non-Darwinian evolution. Science. 1969;**164**(3881):788-798. DOI: 10.1126/science.164.3881.788

[4] Bergeron LA, Besenbacher S, Zheng J, Li P, Bertelsen MF, Quintard B, et al. Evolution of the germline mutation rate across vertebrates. Nature. 2023;**615**(7951):285-291. DOI: 10.1038/s41586-023-05752-y

[5] Lynch M, Ackerman MS, Gout JF, Long H, Way Sung W, Thomas K, et al. Genetic drift, selection and the evolution of the mutation rate. Nature Reviews Genetics. 2016;**17**(11):704-714. DOI: 10.1038/nrg.2016.104

[6] Dańko MJ, Kozłowski J, Vaupel JW, Baudisch A. Mutation accumulation may be a minor force in shaping life history traits. PLoS One. 2012;**7**(4):e34146. DOI: 10.1371/journal.pone.0034146

[7] Wada KN, Doi H, Tanaka S, Wada Y, Furusawa M. A neo-Darwinian algorithm: Asymmetrical mutations due to semiconservative DNA-type replication promote evolution. Proceedings of the National Academy of Sciences of the United States of America. 1993;**90**(24):11934-11938. DOI: 10.1073/pnas.90.24.11934

[8] Watson JD, Crick FH. Molecular structure of nucleic acids: a structure for deoxyribose nucleic acid. Clinical Orthopaedics and Related Research. 2007;**462**:3-5. DOI: 10.1097/BLO.0b013e31814b9304

[9] Spegg V, Altmeyer M. Genome maintenance meets mechanobiology. Review Chromosoma. 2024;**133**(1):15-36. DOI: 10.1007/s00412-023-00807-5

[10] Furusawa M. Okazaki fragment drives evolution. In: OKAZAKI Fragment Memorial Symposium: Celebrating the 50th Anniversary of the Discontinuous DNA Replication Model. 2018. p. 9. Abstract retrieved from Abstracts in the symposium site. Available from: http://bunshi4.bio.nagoya-u.ac.jp/~bunshi4/50okazaki/index.html

[11] Furusawa M. Implications of fidelity difference between the leading and the lagging strand of DNA for the acceleration of evolution. Frontiers in Oncology. 2012;**2**(144):1-10. DOI: 10.3389/fonc.2012.00144

[12] Snedeker J, Wooten M, Chen X. The inherent asymmetry of DNA replication. Annual Review of Cell and Developmental Biology. 2017;**33**(1):291-318. DOI: 10.1146/annurev-cellbio-100616-060447

[13] Sankar TS, Wastuwidyaningtyas BD, Dong Y, Lewis SA, Wang JD. The nature of mutations induced by replication–transcription collisions. Nature. 2016;**535**:178-181. DOI: 10.1038/nature18316

[14] Furusawa M. The disparity mutagenesis model predicts rescue of living things from catastrophic errors. Frontiers in Genetics. 2014;**42**(1):1-8. DOI: 10.3389/fgene.2014.00421

[15] Furusawa M, Doi H. Asymmetrical DNA replication promotes evolution: Disparity theory of evolution. Genetica. 1998;**102-103**:333-347. DOI: 10.1023/A:1017078924245

[16] Domingo E, García-Crespo C, Lobo-Vega R, Perales C. Mutation rates, mutation frequencies, and proofreading-repair activities in RNA virus genetics. Viruses. 2021;**13**(9):1882. Published 2021 Sep 21. DOI: 10.3390/v13091882

[17] Aoki K, Furusawa M. Increase in error threshold for quasi-species by heterogeneous replication accuracy. Physics Review. 2003;**68**:031904-1-031904-6. DOI: 10.1103/PhysRevE.68.031904

[18] Iwaki T, Kawamura A, Ishino Y, Kohno K, Kano Y, Goshima N, et al. Preferential replication-dependent mutagenesis in the lagging DNA strand in *Escherichia coli*. Molecular & General Genetics. 1996;**251**:657-664. DOI: 10.1007/BF02174114

[19] Paul S, Million-Weaver S, Chattopadhyay S, Sokurenko E, Merrikh H. Accelerated gene evolution through replication-transcription conflicts. Nature. 2013;**495**(7442):512-515. DOI: 10.1038/nature11989

[20] Maslowska H, Makiela-Dzbenska K, Mo Y, Schaaper M. High-accuracy lagging-strand DNA replication mediated by DNA polymerase dissociation. PNAS. 2018;**115**(16):4212-4217. DOI: 10.1073/pnas.1720353115

[21] Reijns M, Kemp H, Ding J, de Procé M, Jackson P, Taylor S. Lagging-strand replication shapes the mutational landscape of the genome. Nature. 2015;**518**(7540):502-506. DOI: 10.1038/nature14183

[22] Larrea AA, Lujan SA, McElhinny SAN, Mieczkowski PA, Resnick MA, Gordenin DA, et al. Genome-wide model for the normal eukaryotic DNA replication fork. PNAS. 2010;**107**(41):17674-17679. DOI: 10.1073/pnas.1010178107

[23] Andrianova A, Bazylom A, Nikolaev I, Seplyarskiy B. Human mismatch repair system balances mutation rates between strands by removing more mismatches from the lagging strand. Genome Research. 2017;**27**(8):336-1343. DOI: 10.1101/gr.219915.116

[24] Lujan S, Williams S, Pursell F, Adbulovic-Cui A, Clark B, McElhinny SAN, et al. Mismatch repair balances leading and lagging strand DNA replication fidelity. PLoS Genetics. 2012;**8**(10):e1003016. DOI: 10.1371/journal.pgen.1003016

[25] Lazowski K, Fara M, Vaisman A, Ashton W, Jonczyk P, Fijalkowska J, et al. Strand specificity of ribonucleotide excision repair in *Escherichia coli*. Nucleic Acids Research. 2023;**51**:1766-1782. DOI: 10.1093/nar/gkad038

[26] Koyanagi E, Kakimoto Y, Yoshifuji F, Natsume T, Higashitani A, Ogi T, et al. Global landscape of replicative DNA polymerase usage in the human genome. Nature Communications. 2022;**13**:1-18. DOI: 10.1101/2021.11.14.468503

[27] Baranovskiy G, Babayeva D, Lisova E, Morstadt M, Tahirov H. Structural and functional insight into mismatch extension by human DNA polymerase α. Proceedings of the National Academy of Sciences. 2022;**119**:1-7. DOI: 10.1073/pnas.2111744119

[28] Zhou Z, Lujan S, Burkholder A, Charles J, Dahl J, Farrell C, et al. How asymmetric DNA replication achieves symmetrical fidelity. Nature Structural

& Molecular Biology. 2021;**28**:1020-1028. DOI: 10.1038/s41594-021-00691-6

[29] Dowsett T, Sneeden L, Olson J, McKay-Fleisch J, McAuley E, Scott R, et al. Rate volatility and asymmetric segregation diversify mutation burden in cells with mutator alleles. Communications Biology. 2021;**4**:1-11. DOI: 10.1038/s42003-020-01544-6

[30] Njeri C, Pepenella S, Battapadi T, Bambara A, Balakrishnan L. DNA polymerase delta exhibits altered catalytic properties on lysine acetylation. Genes (Basel). 2023;**14**:791-791. DOI: 10.3390/genes14040774

[31] Kahney W, Ranjan R, Gleason J, Chen X. Symmetry from asymmetry or asymmetry from symmetry? Cold Spring Harbor Symposia on Quantitative Biology. 2017;**82**:305-318. DOI: 10.1101/sqb.2017.82.034272

[32] Ton-Hoang B, Pasternak C, Siguier P, Guynet C, Hickman B, Dyda F, et al. Single-stranded DNA transposition is coupled to host replication. Cell. 2010;**142**(3):398-408. DOI: 10.1016/j.cell.2010.06.034

[33] Seplryarskiy U, Akkuratov E, Akkratova N, Andrianov M, Andrianova M, Nikolaev S, et al. Error-prone bypass of DNA lesions during lagging-strand replication is a common source of germline and cancer mutations. Nature Genetics. 2019;**51**:36-41. DOI: 10.1038/s41588-018-0285-7

[34] Wenger AM, Peluso P, Rowell WJ, Chang PC, Hall RJ, Concepcion GT, et al. Accurate circular consensus long-read sequencing improves variant detection and assembly of a human genome. Nature Biotechnology. 2019;**37**(10):1155-1162. DOI: 10.1038/s41587-019-0217-9. Epub 2019 Aug 12

[35] Singh A. Nanopores for sequencing proteins. Nature Methods. 2023;**20**:1870. DOI: 10.1038/s41592-023-02108-2

[36] Uchimura A, Higuchi M, Minakuchi Y, Ohno M, Toyoda A, Fujiyama A, et al. Germline mutation rates and the long-term phenotypic effects of mutation accumulation in wild type laboratory mice and mutator mice. Genome Research. 2015;**25**:1-10. DOI: 10.1101/gr.186148.114

[37] Inaba S, Sakai H, Kato H, Horiuchi T, Yano H, Ohtsubo Y, et al. Expression of an alcohol dehydrogenase gene in a heterotrophic bacterium induces carbon dioxide-dependent high-yield growth under oligotrophic conditions. Microbiology. 2020;**166**:6. DOI: 10.1099/mic.0.000908

[38] Honma H, Hirai M, Nakamura S, Hakimi H, Kawazu S, Palacpac N, et al. Generation of rodent malaria parasites with a high mutation rate by destructing proofreading activity of DNA polymerase δ. DNA Research. 2014;**21**:439-446. DOI: 10.1093/dnares/dsu009

[39] Yao Z, Wang Q, Dai Z. Recent advances in directed yeast genome evolution. Journal of Fungi (Basel). 2022;**8**(6):635. Published 2022 Jun 15. DOI: 10.3390/jof8060635

[40] Fujihara I, Furusawa M. Disparity mutagenesis model possesses the ability to realize both stable and rapid evolution in response to changing environments without altering mutation rates. Heliyon. 2016;**2**(8):e0014. DOI: 10.1016/j.heliyon.2016.e00141

[41] Fujihara I, Furusawa M. A differential equation, deduced from a DNA-type genetic algorithm with the lagging-strand-biased mutagenesis. Heliyon. 2022;**8**(3):e09155. DOI: 10.1016/j.heliyon.2022.e09155

[42] Akashi M, Fujihara I, Takemura M, Furusawa M. 2-dimensional genetic algorithm exhibited an essentiality of gene interaction for evolution. Journal of Theoretical Biology. 2022;**538**:111044. DOI: 10.1016/j.jtbi.2022.111044

[43] Palazzo F, Kejiou S. Non-Darwinian molecular biology. Frontiers in Genetics. 2022;**13**:831068. Published 2022 Feb 16. DOI: 10.3389/fgene.2022.831068

[44] Robinson J, Kyriazis C, Yuan C, Lohmueller E. Deleterious variation in natural populations and implications for conservation genetics. Annual Review of Animal Biosciences. 2023;**11**:93-114. DOI: 10.1146/annurev-animal-080522-093311

[45] Eigen M. Selforganization of matter and the evolution of biological macromolecules. Die Naturwissenschaften. 1971;**58**(10):465-523. DOI: 10.1007/bf00623322

Chapter 5

NGS and Immunogenetics: Sequencing the HLA Genes

Andreea Mirela Caragea, Laurentiu Camil Bohiltea, Alexandra Constantinescu, Ileana Constantinescu and Radu-Ioan Ursu

Abstract

Next-generation sequencing (NGS) has completely revolutionized the analysis of HLA genes, offering superior resolution and the possibility of identifying previously unknown or rare alleles. NGS technology allows for the complete sequencing of the the HLA locus, the analysis of coding and non-coding regions, and a detailed characterization of haplotypes, with essential benefits in areas such as organ transplantation and in studies of autoimmune diseases. The chapter explores the applications of NGS in personalized medicine, including the identification of neoantigens for oncology immunotherapies and the development of vaccines adapted to the genetic diversity of the population. Bioinformatic and ethical challenges are also discussed. By reducing the limitations of traditional methods and opening up new horizons for research and clinical applications, NGS is redefining the standards in HLA typing and is making a significant contribution to the progress of precision medicine.

Keywords: NGS, data analysis, HLA, immunogenetics, MHC

1. Introduction: The significance of HLA genes in Immunogenetics

The human leukocyte antigen (HLA) genes are part of a complex genetic system known as the major histocompatibility complex (MHC) system and are essential for the functioning of the human body's immune system [1–6]. These genes are located on chromosome 6 and code for proteins that are found on the surface of cells, especially immune cells such as dendritic cells, macrophages, and T lymphocytes [1–3]. These HLA proteins have the role of presenting protein fragments (peptides) from inside the cells, as well as fragments of pathogens that infect the body (bacteria, viruses, and fungi) or tumor cells. Thus, they facilitate the recognition and activation of the immune system when pathogens or abnormal cells are detected.

HLA works through the process of "antigen presentation". This involves the capture of foreign or self-proteins, which are then "broken down" into small fragments (peptides) and bound to HLA proteins [3]. HLA-peptide complexes are presented on the surface of cells for recognition by T cell receptors [2, 3]. T cells have specific receptors that recognize these complexes, and if they identify a foreign fragment, their activation triggers an appropriate immune response, which may include the destruction of infected or abnormal cells [1–3]. This process is essential for distinguishing between "self" and "non-self," allowing the immune system to recognize and eliminate pathogens and defective cells without attacking its own cells [3]. The diversity of HLA genes is crucial to the immune response of the human population. HLA is highly polymorphic, meaning that there are many variants of these genes in the population, each encoding slightly different HLA proteins. This diversity allows the body to recognize a wide range of pathogens, which confers an important evolutionary advantage [3]. Essentially, each individual inherits a unique combination of HLA variants, and their diversity ensures that, within a population, there are always individuals capable of recognizing and fighting different types of pathogens. This can provide collective protection, even when some people have a more vulnerable immune system due to less favorable HLA variants [3].

HLA diversity can also influence a person's susceptibility to certain diseases. For example, some HLA variants are associated with an increased risk of developing autoimmune diseases, such as rheumatoid arthritis, systemic lupus erythematosus, or type 1 diabetes [5–13]. These variants can alter the way the immune system recognizes its own cells and can lead to them being attacked, causing inflammation and tissue damage [13–15]. In contrast, other HLA variants can confer protection against certain infections; for example, some variants are associated with better protection against HIV or hepatitis B infections [1–6].

The interaction between HLA and T cell receptors is essential for the activation of the immune response. T cells recognize peptides presented on HLA molecules using their specific receptors (TCRs). There are two main classes of HLA molecules: class I and class II [1–6]. HLA class I presents peptide fragments originating from within the cell (e.g., viral or tumor proteins), and HLA class II presents peptides originating from external sources, such as bacteria [1–6]. Cytotoxic T cell receptors (CD8+) recognize peptides presented on HLA class I, activating the T cell to destroy infected or cancerous cells, while helper T cell receptors (CD4+) recognize peptides presented on HLA class II, stimulating other components of the immune system, such as B cells or macrophages [1–6]. Thus, by interacting with T cell receptors, HLA plays a crucial role in activating and coordinating the immune response, and the diversity of these genes is essential for the body's effective defense against a wide variety of pathogens.

2. HLA typing: Historical techniques and challenges

Human leukocyte antigen (HLA) typing has evolved significantly over the decades, being essential for the understanding of immune compatibility and for the progress in transplantation and personalized medicine [16]. The history of HLA typing can be divided into several major stages:

1. Discovery of the HLA system (1960–1970): HLA was first identified in the 1960s as part of the major histocompatibility complex (MHC) [1–6]. Early research revealed the importance of this system in graft rejection and in immune mechanisms. The term "HLA" was used to describe molecules on the surface of cells that play a key role in the recognition of "self" and "non-self" by the immune system [16].

2. Serological typing (1970–1990): In the early 1970s, HLA typing techniques were based on serological tests, which involved the use of antibodies to identify HLA

antigens on the surface of leukocytes [1–6]. This was the standard method for determining the HLA types of patients and donors, essential for transplant compatibility [16].

3. PCR and DNA-based technologies (1990–2000): In the 1990s, with the development of polymerase chain reaction (PCR) technology, it became possible to amplify HLA-specific DNA and analyze it at the molecular level [16]. This allowed for much higher resolution and helped identify rare HLA variants, providing better accuracy in HLA typing [1–6].

4. Next-generation sequencing (NGS) (2000–present): NGS technology has completely revolutionized HLA typing, allowing for the complete and detailed sequencing of HLA genes [1–6]. NGS allows for the identification of variants at a much higher resolution than previous technologies, being able to analyze multiple genes simultaneously and discover rare or unexpected variants. These advances have significantly improved disease diagnosis, personalized treatments, and transplant success [16–18].

Thus, HLA typing has evolved from simple serological methods to advanced genomic sequencing technologies, contributing significantly to the understanding of human immunity and to advances in the field of transplantation and personalized medicine.

2.1 Traditional HLA typing models: Serology and PCR

HLA typing, essential for transplantation and immunogenetic studies, was initially performed by traditional methods, such as serological tests and polymerase chain reaction (PCR). These techniques, although revolutionary at the time, have significant limitations that have led to the development of more advanced technologies.

2.1.1 Serological compatibility tests

The serological method, also known as microcytotoxicity, involves the use of specific antibodies to identify HLA antigens on the surface of cells [16]. Cell lysis is observed under a microscope after exposure to antibodies and whole serum, indicating a positive reaction [16].

Advantages: Its simplicity and frequent use in laboratories have made this method a standard for decades.

Limitations: Serological tests offer low resolution, unable to distinguish between genetic variants of HLA alleles. They also depend on the availability of specific antibodies, which can limit accuracy.

2.1.2 PCR-based typing

PCR was a step forward, using specific DNA sequences to identify HLA genes. Methods such as SSP-PCR (specific sequence primer PCR) and SSOP (sequence-specific oligonucleotide probes) have increased the resolution of HLA typing.

Advantages: PCR has allowed for more precise detection of genetic variants and direct analysis of DNA, eliminating the reliance on viable cell samples.

Limitations: PCR methods remain laborious and expensive, requiring advanced equipment and skilled personnel. In addition, the resolution, although higher than serology, is still insufficient to analyze complex variants of HLA alleles.

2.1.3 Challenges and limitations of traditional methods

Low resolution: None of the traditional methods can completely differentiate HLA alleles with similar sequences.

Complexity and cost: The tests require laborious processes and technical expertise, making them inaccessible to smaller laboratories.

Limited reproducibility: Results can vary between laboratories, especially in the case of serological tests, making comparisons difficult [16].

These limitations have led to the adoption of modern technologies such as next-generation sequencing (NGS), which offer high resolution, competitive costs, and detailed analysis of the entire spectrum of HLA variations [16].

2.2 Moving to advanced methods: NGS in HLA typing

With technological progress, traditional HLA typing methods such as serological tests and PCR have been gradually replaced by advanced techniques, the most important of which is next-generation sequencing (NGS) [16–18]. These modern technologies have completely transformed the field of immunogenetics, providing solutions to the limitations and challenges of traditional approaches [16–18].

2.2.1 Limitations of traditional methods and the need for innovation

Serological tests, although useful in the initial context, cannot distinguish between very closely related alleles, limiting the resolution needed in transplantation or research [16]. PCR, although more accurate, also faces similar difficulties in detecting complex variations and requires multiple laborious steps (**Table 1**).

These limitations, such as low resolution, variable reproducibility, and high equipment costs, have created an urgent demand for more robust and efficient methods.

2.2.2 Advantages of NGS in HLA typing

Next-generation sequencing (NGS) has revolutionized genetic analysis with the ability to sequence thousands of DNA fragments simultaneously, providing a complete picture of HLA regions [16, 18].

Aspect	Serological methods	PCR-based techniques	NGS
Resolution	Low	Moderate	High
Throughput	Low	Moderate	High
Allele Detection	Limited to common alleles	Better than serology but limited	Comprehensive, including rare alleles
Cost	Moderate	High	Decreasing
Reproducibility	Variable	Better than serology	High

Table 1.
Comparison of HLA typing techniques.

1. High resolution: Unlike traditional methods, NGS allows for ultra-detailed typing, precisely identifying alleles variants (at the allele level) of HLA genes. This is essential in transplantation, where an exact HLA match can prevent graft rejection [16].

2. Complete coverage: NGS analyzes the entire HLA locus, including non-coding regions, providing detailed genetic information about the structure of haplotypes [18].

3. Efficiency: Processing multiple samples simultaneously reduces analysis time and costs per sample, making NGS a scalable solution [18].

4. Reliability and reproducibility: Unlike laborious manual methods, NGS produces standardized data, minimizing interlaboratory variation [18].

2.2.3 Impact of NGS on HLA typing

With NGS, researchers can address complex questions about HLA variability and its relationship to autoimmune diseases, susceptibility to infections, and transplantation. In addition, the high resolution allows for the discovery of rare or novel HLA alleles that were previously undetectable [18].

In conclusion, the transition from traditional methods to NGS is not just a technological evolution, but a fundamental change that redefines the boundaries of immunogenetics research. With its combination of sensitivity, precision, and efficiency, NGS has become the gold standard in HLA typing.

3. The rise of next-generation sequencing (NGS)

Next-generation sequencing (NGS) is an advanced technology that allows simultaneous analysis of large DNA sequences, providing a detailed understanding of the genome [19]. Unlike traditional sequencing methods, such as the Sanger method, which sequence small fragments and require a long time, NGS is fast, scalable, and highly accurate.

3.1 How does NGS work?

NGS uses parallel sequencing techniques, in which millions of DNA fragments are read simultaneously. This capability allows the reconstruction of complex regions, such as HLA genes, which are difficult to analyze by traditional methods.

3.1.1 Benefits of NGS

1. High resolution: NGS can identify genetic variations at the allele level, including rare polymorphisms or non-coding regions, providing a complete picture of DNA.

2. Processing speed: NGS can analyze thousands of fragments simultaneously, reducing the time required for sequencing compared to classical methods.

3. Multiplexing: This technology allows the analysis of multiple genes in parallel, which is essential for complex genetic studies, such as those involving multiple HLA loci.

4. Scalability: NGS is efficient in processing a large volume of samples, making it ideal for research and clinical applications.

3.1.2 Applications in human genetics

NGS has had a profound impact on genetic research, especially in the analysis of the HLA system, where detailed resolution is crucial. In immunogenetics, NGS allows precise HLA typing, which is essential in transplantation, where exact matching reduces the risk of graft rejection [19]. In addition, NGS facilitates the study of autoimmune diseases, susceptibility to infections, and genetic diversity in different populations.

In conclusion, NGS has revolutionized genetic analysis through its efficiency, precision, and ability to explore genomic variability. Its applications, especially in the study of HLA, continue to advance personalized medicine and the understanding of complex diseases [19].

4. Advances in NGS: Unlocking the complexity of HLA genes

Human leukocyte antigen (HLA) genes are essential for the functioning of the adaptive immune system, presenting pathogen-derived peptides to T-cell receptors [6]. These genes are highly polymorphic, reflecting an evolutionary adaptation to diverse selective pressures, such as exposure to specific pathogens. Detailed sequencing of HLA genes is crucial in transplantation, immunotherapy, and understanding susceptibility to autoimmune and infectious diseases [6].

4.1 HLA gene sequencing: Deciphering diversity with NGS

Next-generation sequencing (NGS) is revolutionizing HLA gene analysis with its ability to identify allelic variants (haplotypes) and rare polymorphisms. Unlike traditional methods, which rely on specific probes or targeted sequences, NGS uses massively parallel sequencing to generate comprehensive data, including intronic and non-coding regions [19]. This is essential because many HLA alleles differ by single substitutions or small insertions and deletions, which can have a significant impact on protein function.

By using advanced bioinformatic algorithms, NGS can resolve the ambiguity problems caused by the high similarities between HLA genes and their pseudogenes [19, 20]. Thus, the technology allows for precise characterization of haplotypes, including rare or novel alleles, which are critical for transplant compatibility and population studies.

4.2 Complexity of HLA genes and the resolution provided by NGS

The complexity of HLA genes stems from their extreme polymorphism: Each gene can have hundreds of functional alleles. For example, HLA-B has over 5000 known alleles, and this number continues to grow [21]. This diversity plays a key role in the recognition of specific antigens and in determining variable immunological responses between individuals [21].

NGS allows for nucleotide-level resolution, identifying variations even in structurally very similar regions. Furthermore, the ability to analyze the entire HLA locus, including regulatory regions, allows for a more comprehensive understanding of the impact of these genes on immunological and clinical phenotypes [19].

4.3 Clinical applications of detailed HLA sequencing

In the field of transplantation, detailed analysis of HLA genes with NGS ensures perfect compatibility between donor and recipient, reducing the risk of acute or chronic graft rejection [22]. In addition, it allows the identification of the minimal immunogenicity required for improved immunological tolerance.

In oncology, HLA profiling plays a crucial role in the selection of patients for immunotherapies, such as those using immune checkpoint inhibitors [23–27]. NGS-facilitated HLA studies are also essential in the development of personalized vaccines based on specific tumor neoantigens.

In addition to these applications, HLA analysis performed by NGS improves research on autoimmune diseases (e.g., the association of HLA-B27 with ankylosing spondylitis) and susceptibility to infections (such as the link between HLA and HIV) [23–29]. The high resolution and processing speed offered by NGS are transforming the understanding and applicability of HLA in medicine, opening new perspectives for diagnosis, treatment, and prevention.

5. Bioinformatics in HLA data interpretation

5.1 NGS data processing in HLA sequencing

The processing of raw data obtained by next-generation sequencing (NGS) for HLA gene analysis involves several crucial steps, each essential for obtaining an accurate profile (**Figure 1**).

1. Raw sequence acquisition: NGS generates short or long reads of DNA sequences, which include the coding and non-coding regions of HLA genes. The data are collected as FASTQ files containing the DNA sequences and quality scores [30–33].

2. Data preprocessing: This step involves filtering and cleaning the reads to remove artifacts, sequences with low quality scores, or contaminations. Specific tools, such as Trimmomatic or FastQC, are used for quality assessment and optimization of the reads [30–33].

3. Read alignment: The reads are mapped to a reference sequence of the human genome using alignment algorithms, such as BWA-MEM or Bowtie2. For HLA, alignment is more complex due to the high homologies between HLA alleles and pseudogenes [30–33].

4. HLA haplotype reconstruction: The aligned data is processed to identify haplotypes specific to each individual, using databases such as IMGT/HLA [34]. This step requires specialized algorithms for the analysis of genes with high polymorphism [34].

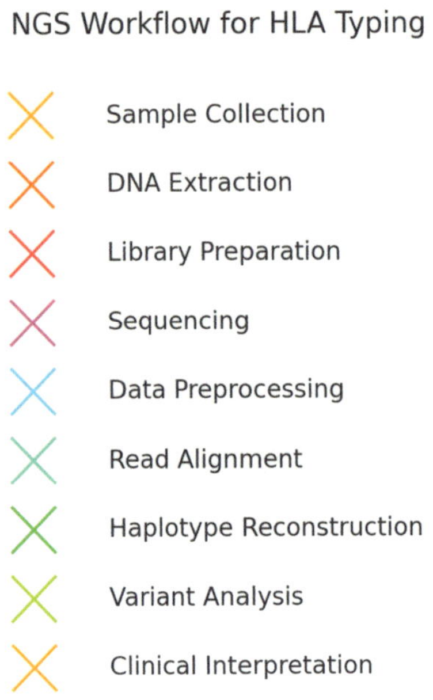

Figure 1.
NGS workflow for HLA typing.

5. Variant identification: All-cell variations (SNPs, insertions/deletions) are detected and associated with HLA alleles using advanced bioinformatics tools.

5.2 Algorithms and software used in HLA analysis by NGS

1. OptiType: An accurate algorithm for HLA typing based on NGS data, using Bayesian methods to reconstruct possible haplotypes [35].

2. HLA-HD: Optimized software for NGS data analysis, capable of identifying complex haplotypes and detecting rare variants [30].

3. Seq2HLA: An RNA-Seq-based program that allows expression analysis and HLA typing [36, 37].

4. ArcasHLA: Advanced HLA typing algorithm that integrates raw data with HLA databases for fast and accurate identification [38].

5. SAMtools/BCFtools: General tools for variant analysis, useful in calibrating HLA alleles [39].

5.3 Bioinformatics challenges in HLA analysis

1. Complexity of HLA genes: HLA genes exhibit high homologies between different loci and pseudogenes, which can lead to misalignments and ambiguities in haplotype identification [34].

2. Variant calibration: Detection of rare alleles and subtle differences between alleles requires precise calibration, and technical variations in NGS data can complicate this process.

3. Large volumes of data: NGS generates a massive amount of data that requires robust computational infrastructure for storage and analysis [40].

4. Database limitations: HLA databases (such as IMGT/HLA) are continuously updated, but do not include all possible variants, which may limit complete identification [34].

5. Computational costs: HLA analysis requires specialized bioinformatics algorithms, which involve high computational power and associated costs.

Despite these challenges, advances in bioinformatics algorithms and the continuous updating of HLA databases allow for increasingly accurate and efficient interpretation of NGS data, contributing to clinical and research applicability.

6. Clinical applications of HLA sequencing: From transplantation to autoimmune diseases

See **Table 2**.

Application	Importance of HLA typing
Organ Transplantation	Ensures donor–recipient compatibility, reducing rejection risk
Autoimmune Disease Diagnosis	Identifies genetic predisposition to diseases like lupus, rheumatoid arthritis
Personalized Medicine	Customizes therapies based on HLA profile, minimizing adverse drug reactions
Vaccine Development	Designs vaccines targeting diverse genetic populations for robust responses

Table 2.
Clinical applications of HLA sequencing.

6.1 Transplants: The role of HLA sequencing in donor–recipient compatibility

HLA sequencing plays a critical role in determining immunological compatibility between donor and recipient in organ, tissue, and bone marrow transplantation [1, 2]. HLA genes are responsible for presenting cellular antigens to the immune system, so significant differences between the HLA alleles of the donor and recipient can trigger an aggressive immune response, resulting in transplant rejection [1, 2].

Advanced technologies, such as NGS sequencing, allow for the high-resolution identification of HLA alleles, including rare variants. This level of detail is crucial for selecting compatible donors and minimizing the risk of immune reactions, such as graft-versus-host disease (GvHD) in bone marrow transplantation. By using HLA sequencing, the success rate of transplants can be significantly increased, and the duration of graft survival can be prolonged.

6.2 Autoimmune diseases: Understanding genetic predisposition through HLA sequencing

HLA sequencing provides valuable insights into genetic susceptibility to autoimmune diseases such as systemic lupus erythematosus, rheumatoid arthritis, and type 1 diabetes [6–10]. Many autoimmune diseases are closely associated with specific HLA alleles. For example:

HLA-DR4 is frequently associated with rheumatoid arthritis.

HLA-DQ8 and HLA-DQ2 are implicated in celiac disease.

NGS helps identify alleles involved in these diseases with high accuracy, contributing to understanding the molecular mechanisms underlying autoimmunity. This may facilitate early diagnosis and the development of targeted treatments.

6.3 Personalized medicine: Personalizing treatments through HLA analysis

HLA sequencing is an essential component of personalized medicine, especially in the context of immunotherapy and vaccine development [41–43]. By analyzing patients' HLA profiles in detail:

Cancer immunotherapy: HLA sequencing can guide the selection of neoantigens for T-cell-based therapies, ensuring a tumor-specific immune response.

Adverse drug reactions: Some HLA alleles, such as HLA-B*57:01, are associated with severe drug reactions, such as hypersensitivity to abacavir [43]. Identifying these alleles allows doctors to adjust treatments to minimize risks.

Vaccine development: Understanding HLA diversity helps create vaccines that can generate robust immune responses in diverse populations [44].

Thus, HLA sequencing is a cornerstone in targeting treatments and optimizing medical care based on individual genetic characteristics.

7. The future of HLA sequencing: Personalized medicine and beyond

The future of HLA sequencing is closely linked to technological innovations and the integration of genetic data in personalized medicine. Advances in NGS will allow for increasingly higher resolution and faster and more accessible analysis of the complex diversity of HLA genes [45]. This will facilitate the development of personalized treatments, optimization of transplant compatibility, and a better understanding of genetic predispositions to autoimmune or infectious diseases.

At the same time, the integration of artificial intelligence and advanced bioinformatics will revolutionize the interpretation of HLA data, paving the way for personalized immuno-oncology therapies and vaccines tailored to human genetic diversity [46]. In the long term, full genome sequencing, including HLA analysis, will become standard practice, providing clearer insights into individual and collective health.

7.1 Personalized medicine: The revolution brought by HLA sequencing

HLA sequencing technology, especially through next-generation sequencing (NGS), plays a central role in the transformation of personalized medicine. By analyzing a patient's HLA profile in detail, doctors can tailor treatments to address their unique genetic characteristics. For example, cancer immunotherapy, which relies on stimulating the immune system to recognize and eliminate tumor cells, benefits enormously from the information provided by HLA sequencing [45]. This allows the identification of tumor-specific neoantigens, which are then used to personalize therapeutic vaccines or T-cell therapies.

In addition, HLA sequencing helps prevent adverse drug reactions. For example, the HLA-B*57:01 allele is associated with severe reactions to abacavir, a drug used to treat HIV [43]. Early identification of this allele ensures that safer treatment is chosen for the patient.

7.2 Whole genome sequencing: A new frontier in personalized medicine

As technologies advance, whole genome analysis is becoming increasingly accessible. Unlike targeted HLA gene sequencing, whole genome analysis provides a holistic picture of a patient's genetic predispositions. This allows the identification of variants associated not only with the immune response, but also with susceptibility to other chronic diseases, such as diabetes, cardiovascular disease, and cancer [45].

By integrating HLA sequencing with complete genomic data, researchers can develop new diagnostic and treatment strategies. This approach can detect hidden genetic risks and guide early interventions, ensuring more effective prevention.

7.3 Future possibilities in treatments: The impact of NGS and HLA on tomorrow's medicine

NGS technology, combined with detailed HLA analysis, will open up new opportunities in the treatment of complex diseases. In oncology, the discovery of HLA-associated neoantigens will facilitate the development of more effective therapies, such as personalized vaccines and CAR-T therapies [45, 47].

In the case of rare diseases, where genetic diversity plays a major role, NGS can identify HLA variants that contribute to the manifestation of pathology, thus providing targets for new treatments. Also, in infectious diseases, HLA analysis can guide the development of vaccines tailored to the genetic diversity of global populations.

Thus, the future of medical treatments will be increasingly guided by advanced genomics, revolutionizing the way doctors approach the prevention, diagnosis, and treatment of diseases.

8. NGS in HLA research: New horizons in immunogenetics

Next-generation sequencing (NGS) has revolutionized HLA research, enabling detailed, high-resolution analysis of HLA gene variability (**Figure 2**).

NGS facilitates the identification and characterization of rare HLA gene variants that were not detectable by traditional typing methods. This advanced technology contributes to the understanding of HLA diversity in diverse populations, allowing researchers to explore how these variants influence disease susceptibility and immune responses. In addition, NGS allows the study of complex interactions between HLA and pathogens, providing valuable information for the development of personalized therapies and vaccines. This approach opens up new possibilities for precision medicine and improved treatments, including in the fields of transplantation and autoimmune diseases (**Figure 1**).

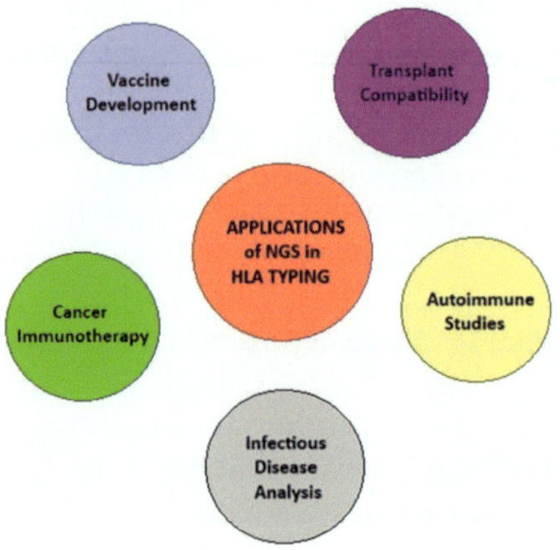

Figure 2.
Applications of NGS in HLA typing.

8.1 The new frontier of HLA research

Next-generation sequencing (NGS) is redefining the exploration of HLA variability, providing an unprecedented level of detail. In population studies, NGS allows for the precise analysis of rare and common HLA alleles, contributing to the understanding of genetic adaptations and selective pressures throughout human history. This opens new perspectives on HLA diversity in isolated or under-represented populations, with applications in precision medicine and in studies of migration and evolution.

8.2 Immunogenetics of the future

Future research, facilitated by NGS, will deepen the understanding of immunological mechanisms by comprehensively analyzing HLA regions and their interactions with other genes. Identification of genetic variations associated with autoimmune and infectious diseases will contribute to prevention and personalized therapies [48]. Furthermore, emerging technologies can integrate HLA analysis with other omics data (transcriptomics and proteomics), building a more complete picture of the immune response [48].

8.3 HLA in interaction with pathogens

The HLA system plays a key role in the immune response by presenting pathogenic antigens. Recent studies, supported by NGS, analyze how certain HLA alleles influence resistance or susceptibility to infections such as malaria, HIV, or tuberculosis [27, 29, 49, 50]. This technology allows the rapid identification of HLA variants involved in effective responses to specific pathogens. These discoveries are fundamental for the development of vaccines and treatments tailored to individual genetic diversity, contributing to the fight against emerging and pandemic diseases [29].

9. Ethical and practical considerations in HLA sequencing

Ethical issues in HLA typing include issues such as confidentiality and use of genetic data, informed consent, and genetic discrimination [51]. Because HLA information can reveal genetic predispositions to autoimmune diseases or risks to transplant success, there is a risk that it could be misused by employers or insurance companies (**Table 3**).

In addition, the typing process must be carried out with the patient's clear consent, ensuring that they understand how their genetic data will be used. It is also important that access to these advanced technologies is equitable and does not discriminate against certain population groups.

9.1 Confidentiality of genetic data

The confidentiality of genetic data obtained through HLA sequencing is a major concern, given its highly personal nature [51]. Although this information can bring significant benefits for personalized diagnosis and treatment, the risks associated with its disclosure may include genetic discrimination and violation of individual privacy. Thus, there is a need for strict regulations and clear data protection policies, including anonymization and encryption. Protecting patient rights and ensuring informed consent are also essential to prevent the misuse of genetic information, especially in the fields of insurance and employment.

9.2 Equity in access to technology

Access to advanced sequencing technologies, including NGS for HLA analysis, remains an important topic of discussion, especially in resource-limited regions. While these technologies have the potential to revolutionize genetic diagnosis and treatment, the high costs and required technological infrastructure can be barriers for many populations. Lack of access to advanced sequencing can contribute to disparities in health care and disease prevention, and it is essential that governments and international organizations promote equity in access to these innovative technologies.

9.3 Impact of medical decisions

HLA analysis plays a crucial role in medical decision-making, especially in the fields of transplantation, autoimmune diseases, and personalized medicine. However, errors in the interpretation of genetic data or wrong predictions can lead to incorrect

Issue	Explanation
Privacy and Confidentiality	Risks of misuse of sensitive genetic data in employment or insurance
Informed consent	Ensuring participants understand usage and implications of genetic information
Equity in access	Disparity in availability of advanced technology in low-resource settings
Genetic Discrimination	Potential misuse of genetic predisposition information by third parties

Table 3.
Ethical considerations in HLA sequencing.

diagnoses or wrong therapeutic decisions, directly affecting the patient's health. For example, incorrect HLA typing can lead to the rejection of an organ transplant or the administration of a treatment that is not optimal for the patient. Thus, it is necessary that HLA analysis is supported by well-trained professionals and is accompanied by a rigorous validation system to minimize the associated risks.

10. Conclusions

In conclusion, NGS represents a revolution in HLA typing and will continue to deepen our understanding of human genetics and immune interactions, providing new insights and solutions for a wide range of clinical and research applications.

We can draw several important conclusions regarding the evolution and impact of HLA typing technologies, especially in the context of next-generation sequencing (NGS) technologies.

1. Evolution of HLA typing technologies: From the first serological techniques, which were limited in resolution, to the advent of PCR and, more recently, next-generation sequencing (NGS), progress in HLA typing has been essential for the success of transplants, the understanding of autoimmune diseases, and the development of personalized medicine [16, 19]. NGS represents a major leap forward, allowing detailed analysis of genetic variants and the identification of rare variants, with a significant impact on diagnosis and treatment.

2. Impact of NGS on research and clinical applications: NGS technology has revolutionized immunogenetics research, allowing a much more detailed understanding of HLA variability at the population and individual level [16, 19]. This contributes not only to the identification of compatibility in transplants, but also to a better understanding of predispositions to autoimmune, infectious diseases and cancer, providing an essential tool for personalized medicine.

3. Traditional challenges and limitations: Despite advances, traditional HLA typing methods, such as serological tests and PCR, had significant limitations, including low resolution, high cost, and technological complexity. NGS has overcome these challenges, but this technology also comes with its own challenges, including the need for complex bioinformatic algorithms and the correct interpretation of genetic variants [16, 18, 19].

4. The future of HLA sequencing and its applications: As NGS technology continues to evolve, its potential to transform personalized medicine is immense. Whole genome sequencing and detailed analysis of HLA profiles will play a key role in the development of more precise treatments, including in the fields of cancer, rare diseases, and vaccinology [48]. At the same time, NGS will support research on HLA-pathogen interactions, providing vital information for the development of more effective therapies.

5. Ethical and social challenges: Despite technological advances, important challenges remain related to the privacy of genetic data, equitable access to advanced technologies, and the potential impact of medical decisions influenced by HLA analysis. Ensuring a robust and equitable ethical framework will be essential as these technologies become more accessible [50, 51].

Conflict of interest

The authors declare no conflict of interest.

Author details

Andreea Mirela Caragea[1,2], Laurentiu Camil Bohiltea[3], Alexandra Constantinescu[1], Ileana Constantinescu[1,2] and Radu-Ioan Ursu[3]*

1 Carol Davila University of Medicine and Pharmacy, Bucharest, Romania

2 Center for Immunogenetics and Virology, Fundeni Clinical Institute, Bucharest, Romania

3 Department of Medical Genetics, Carol Davila University of Medicine and Pharmacy, Bucharest, Romania

*Address all correspondence to: dr.radu.ursu@gmail.com

References

[1] Caragea AM, Ursu RI, Maruntelu I, Tizu M, Constantinescu AE, Tălăngescu A, et al. High resolution HLA-A, HLA-B, and HLA-C allele frequencies in Romanian hematopoietic stem cell donors. International Journal of Molecular Sciences. 2024;**25**(16):8837. DOI: 10.3390/ijms25168837ences

[2] Caragea MA, Ursu IR, Visan DL, Maruntelu I, Iordache P, Constantinescu A, et al. High-resolution HLA-DRB1 allele frequencies in a Romanian cohort of stem cell donors. Balkan Journal of Medical Genetics: BJMG. 2024;**27**(1):43-49. DOI: 10.2478/bjmg-2024-0009

[3] Delves PJ, Martin SJ, Burton DR, Roitt IM. Part. 1. Fundamentals of immunology. In: Martin DPJ, editor. Roitt's Essential Immunology. 13th ed. Oxford, UK: John Wiley and Sons, Ltd.; 2017. pp. 1-291

[4] Gabriel C, Fürst D, Faé I, Wenda S, Zollikofer C, Mytilineos J, et al. HLA typing by next-generation sequencing—Getting closer to reality. Tissue Antigens. 2014;**83**(2):65-75. DOI: 10.1111/tan.12298

[5] Rich RR, Fleisher TA, Shearer WT, Schroeder HW Jr, Frew AJ, Weyand CM. Part one: Principles of immune response. In: Rich RR, editor. Clinical Immunology. Principles and Practice. 4th ed. UK: Elsevier Saunders; 2013. pp. 3-183

[6] Kochi Y, Suzuki A, Yamada R, Yamamoto K. Genetics of rheumatoidarthritis: Underlying evidence of ethnic differences. Journal of Autoimmunity. 2009;**32**:158-162

[7] Newton JL, Harney SM, Wordsworth BP, Brown MA. A review of the MHC genetics of rheumatoid arthritis. Genes and Immunity. 2004;**5**:151-157

[8] Salvarani C, Macchioni PL, Mantovani W, Bragliani M, Collina E, Cremonesi T, et al. HLA-DRB1 alleles associated with rheumatoid arthritis in northern Italy: Correlation with disease severity. British Journal of Rheumatology. 1998;**37**:165-169

[9] Bridges SL Jr, Kelley JM, Hughes LB. The HLA-DRB1 shared epitope in caucasians with rheumatoid arthritis: A lesson learned from tic-tactoe. Arthritis and Rheumatism. 2008;**58**:1211-1215

[10] Tuokko J, Nejentsev S, Luukkainen R, Toivanen A, Ilonen J. HLA haplotype analysis in Finnish patients with rheumatoid arthritis. Arthritis and Rheumatism. 2001;**44**:315-322

[11] Lee HS, Lee KW, Song GG, Kim HA, Kim SY, Bae SC. Increased susceptibility to rheumatoid arthritis in Koreans heterozygous for HLA-DRB1*0405 and *0901. Arthritis and Rheumatism. 2004;**50**:3468-3475

[12] Delgado-Vega AM, Anaya JM. Meta-analysis of HLA DRB polymorphism in Latin American patients with rheumatoid arthritis. Autoimmunity Reviews. 2007;**6**:402-408

[13] Seidl C, Koch U, Buhleier T, Frank R, Möller B, Markert E, et al. HLA-DRB1*04 subtypes are associated with increased inflammatory activity in early rheumatoid arthritis. British Journal of Rheumatology. 1997;**36**:941-944

[14] Kinikli G, Ateş A, Turgay M, Akay G, Kinikli S, Tokgöz G. HLA-DRB1 genes and disease severity in rheumatoid

arthritis in Turkey. Scandinavian Journal of Rheumatology. 2003;**32**:277-280

[15] Uçar F, Karkucak M, Alemdaroğlu E, Capkin E, Yücel B, Sönmez M, et al. HLADRB1 allele distribution and its relation to rheumatoid arthritis in eastern Black Sea Turkish population. Rheumatology International. 2012;**32**:1003-1007

[16] Edgerly CH, Weimer ET. The past, present, and future of HLA typing in transplantation. Methods in Molecular Biology. 2018;**1802**:1-10. DOI: 10.1007/978-1-4939-8546-3_1

[17] Zhou Y, Song L, Li H. Full resolution HLA and KIR genes annotation for human genome assemblies. bioRxiv. 2024;**34**(11):1931-1941. DOI: 10.1101/gr.278985.124

[18] Liu P, Yao M, Gong Y, Song Y, Chen Y, Ye Y, et al. Benchmarking the human leukocyte antigen typing performance of three assays and seven next-generation sequencing-based algorithms. Frontiers in Immunology. 2021;**12**:652258. DOI: 10.3389/fimmu.2021.652258

[19] Profaizer T, Kumánovics A. Human leukocyte antigen typing by next-generation sequencing. Clinics in Laboratory Medicine. 2018;**38**(4):565-578. DOI: 10.1016/j.cll.2018.07.006. Epub 2018 Oct 5

[20] Lai SK, Luo AC, Chiu IH, Chuang HW, Chou TH, Hung TK, et al. A novel framework for human leukocyte antigen (HLA) genotyping using probe capture-based targeted next-generation sequencing and computational analysis. Computational and Structural Biotechnology Journal. 2024;**23**:1562-1571. DOI: 10.1016/j.csbj.2024.03.030

[21] Olson E, Geng J, Raghavan M. Polymorphisms of HLA-B: Influences on assembly and immunity. Current Opinion in Immunology. 2020;**64**:137-145. DOI: 10.1016/j.coi.2020.05.008. Epub 2020 Jun 30

[22] Boegel S, Löwer M, Bukur T, Sahin U, Castle JC. A catalog of HLA type, HLA expression, and neo-epitope candidates in human cancer cell lines. Oncoimmunology. 2014;**3**(8):e954893. DOI: 10.4161/21624011.2014.954893

[23] Wajda A, Sivitskaya L, Paradowska-Gorycka A. Application of NGS Technology in Understanding the pathology of autoimmune diseases. Journal of Clinical Medicine. 2021;**10**(15):3334. DOI: 10.3390/jcm10153334

[24] Muehling LM, Mai DT, Kwok WW, Heymann PW, Pomés A, Woodfolk JA. Circulating memory CD4+ T cells target conserved epitopes of rhinovirus capsid proteins and respond rapidly to experimental infection in humans. Journal of Immunology. 2016;**197**(8):3214-3224. DOI: 10.4049/jimmunol.1600663

[25] Faner R, James E, Huston L, Pujol-Borrel R, Kwok WW, Juan M. Reassessing the role of HLA-DRB3 T-cell responses: Evidence for significant expression and complementary antigen presentation. European Journal of Immunology. 2010;**40**(1):91-102. DOI: 10.1002/eji.200939225

[26] Becerra-Artiles A, Cruz J, Leszyk JD, Sidney J, Sette A, Shaffer SA, et al. Naturally processed HLA-DR3-restricted HHV-6B peptides are recognized broadly with polyfunctional and cytotoxic CD4 T-cell responses. European Journal of Immunology. 2019;**49**(8):1167-1185. DOI: 10.1002/eji.201948126

[27] Anguille S, Fujiki F, Smits EL, Oji Y, Lion E, Oka Y, et al. Identification of a

Wilms' tumor 1-derived immunogenic CD4(+) T-cell epitope that is recognized in the context of common Caucasian HLA-DR haplotypes. Leukemia. 2013;**27**(3):748-750. DOI: 10.1038/leu.2012.248

[28] Yossef R, Tran E, Deniger DC, Gros A, Pasetto A, Parkhurst MR, et al. Enhanced detection of neoantigen-reactive T cells targeting unique and shared oncogenes for personalized cancer immunotherapy. JCI Insight. 2018;**3**(19):e122467. DOI: 10.1172/jci.insight.122467

[29] Galperin M, Farenc C, Mukhopadhyay M, Jayasinghe D, Decroos A, Benati D, et al. CD4+ T cell-mediated HLA class II cross-restriction in HIV controllers. Science Immunology. 2018;**3**(24):eaat0687. DOI: 10.1126/sciimmunol.aat0687

[30] Kawaguchi S, Higasa K, Shimizu M, Yamada R, Matsuda F. HLA-HD: An accurate HLA typing algorithm for next-generation sequencing data. Human Mutation. 2017;**38**(7):788-797. DOI: 10.1002/humu.23230. Epub 2017 May 12

[31] Mei S, Li F, Leier A, Marquez-Lago TT, Giam K, Croft NP, et al. A comprehensive review and performance evaluation of bioinformatics tools for HLA class I peptide-binding prediction. Briefings in Bioinformatics. 2020;**21**(4):1119-1135. DOI: 10.1093/bib/bbz051

[32] Wu G, Xiao G, Yan Y, Guo C, Hu N, Shen S. Bioinformatics analysis of the clinical significance of HLA class II in breast cancer. Medicine (Baltimore). 2022;**101**(40):e31071. DOI: 10.1097/MD.0000000000031071

[33] Usureau C, Jacob V, Dubois V, Masson D, Jollet I, Desoutter J, et al. HLA graph, a free and ready-to-use bioinformatics tool to explore anti-HLA eplets reactivity pattern. HLA. 2022;**100**(3):244-253. DOI: 10.1111/tan.14701 E. Epub 2022 Jun 18

[34] Robinson J, Barker DJ, Georgiou X, Cooper MA, Flicek P, Marsh SGE. IPD-IMGT/HLA Database. Nucleic Acids Research. 2020;**48**(D1):D948-D955. DOI: 10.1093/nar/gkz950

[35] Szolek A, Schubert B, Mohr C, Sturm M, Feldhahn M, Kohlbacher O. OptiType: Precision HLA typing from next-generation sequencing data. Bioinformatics. 2014;**30**(23):3310-3316. DOI: 10.1093/bioinformatics/btu548. Epub 2014 Aug 20

[36] Boegel S, Löwer M, Schäfer M, Bukur T, de Graaf J, Boisguérin V, et al. HLA typing from RNA-Seq sequence reads. Genome Medicine. 2012;**4**(12):102. DOI: 10.1186/gm403

[37] Boegel S, Scholtalbers J, Löwer M, Sahin U, Castle JC. In silico HLA typing using standard RNA-Seq sequence reads. Methods in Molecular Biology. 2015;**1310**:247-258. DOI: 10.1007/978-1-4939-2690-9_20

[38] Orenbuch R, Filip I, Comito D, Shaman J, Pe'er I, Rabadan R. arcasHLA: High-resolution HLA typing from RNAseq. Bioinformatics. 2020;**36**(1):33-40. DOI: 10.1093/bioinformatics/btz474

[39] Danecek P, Bonfield JK, Liddle J, Marshall J, Ohan V, Pollard MO, et al. Twelve years of SAMtools and BCFtools. GigaScience. 2021;**10**(2):giab008. DOI: 10.1093/gigascience/giab008

[40] Naranbhai V, Viard M, Dean M, Groha S, Braun DA, Labaki C, et al. HLA-A*03 and response to immune checkpoint blockade in cancer: An epidemiological biomarker study. The Lancet Oncology. 2022;**23**(1):172-184. DOI: 10.1016/

S1470-2045(21)00582-9. Epub 2021 Dec 9

[41] International multiple sclerosis genetics consortium, Wellcome Trust case control consortium 2. In: Sawcer S, Hellenthal G, Pirinen M, Spencer CC, Patsopoulos NA, Moutsianas L et al. Genetic risk and a primary role for cell-mediated immune mechanisms in multiple sclerosis. Nature. 2011;**476**(7359):214-219. DOI: 10.1038/nature10251

[42] Ursu RI, Cucu N, Ursu GF, et al. Frequency study of the FTO and ADRB3 genotypes in a Romanian cohort of obese children. Romanian Biotechnology Letters. 2016;**21**(3):11610-11620

[43] Dean L. Abacavir therapy and *HLA-B*57:01* genotype. In: Pratt VM, Scott SA, Pirmohamed M, Esquivel B, Kattman BL, Malheiro AJ, editors. Medical Genetics Summaries [Internet]. Bethesda (MD): National Center for Biotechnology Information (US); 2015; 2012–

[44] Zhao L, Zhang M, Cong H. Advances in the study of HLA-restricted epitope vaccines. Human Vaccines & Immunotherapeutics. 2013;**9**(12):2566-2577. DOI: 10.4161/hv.26088. Epub 2013 Aug 16

[45] Misra MK, Mostafa A, Charron D. Editorial: HLA in personalized medicine. Frontiers in Genetics. 2024;**15**:1480936. DOI: 10.3389/fgene.2024.1480936

[46] Lu Z, Zou Q, Wang M, Han X, Shi X, Wu S, et al. Artificial intelligence improves the diagnosis of human leukocyte antigen (HLA)-B27-negative axial spondyloarthritis based on multi-sequence magnetic resonance imaging and clinical features. Quantitative Imaging in Medicine and Surgery. 2024;**14**(8):5845-5860. DOI: 10.21037/qims-24-729. Epub 2024 Jul 30

[47] Chen X, Tan B, Xing H, Zhao X, Ping Y, Zhang Z, et al. Allogeneic CAR-T cells with of HLA-A/B and TRAC disruption exhibit promising antitumor capacity against B cell malignancies. Cancer Immunology, Immunotherapy. 2024;**73**(1):13. DOI: 10.1007/s00262-023-03586-1

[48] Cornaby C, Weimer ET. HLA typing by next-generation sequencing: Lessons learned and future applications. Clinics in Laboratory Medicine. 2022;**42**(4):603-612. DOI: 10.1016/j.cll.2022.09.013

[49] Avila-Rios S, Carlson JM, John M, Mallal S, Brumme ZL. Clinical and evolutionary consequences of HIV adaptation to HLA: Implications for vaccine and cure. Current Opinion in HIV and AIDS. 2019;**14**(3):194-204. DOI: 10.1097/COH.0000000000000541

[50] Dawkins BA, Garman L, Cejda N, Pezant N, Rasmussen A, Rybicki BA, et al. Novel HLA associations with outcomes of mycobacterium tuberculosis exposure and sarcoidosis in individuals of African ancestry using nearest-neighbor feature selection. Genetic Epidemiology. 2022;**46**(7):463-474. DOI: 10.1002/gepi.22490. Epub 2022

[51] Pennings G, Schots R, Liebaers I. Ethical considerations on preimplantation genetic diagnosis for HLA typing to match a future child as a donor of haematopoietic stem cells to a sibling. Human Reproduction. 2002;**17**(3):534-538. DOI: 10.1093/humrep/17.3.534